Environmental Chemistry of Herbicides

Volume I

Editor

R. Grover

Section Head
Environmental Chemistry of Herbicides Section
Research Branch, Agriculture Canada,
Research Station, Regina
Saskatchewan, Canada

CRC Press, Inc.
Boca Raton, Florida

Library of Congress Cataloging-in-Publication Data

Environmental chemistry of herbicides.

  Bibliography: p.
  Includes index.
  1. Herbicides. 2. Herbicides--Environmental aspects.
3. Soils--Herbicide content. 4. Environmental
chemistry. I. Grover, R. (Raj)
SB951.4.E58 1988   574.5'222   87-28975
ISBN 0-8493-4375-5 (set)
ISBN 0-8493-4376-3 (v. 1)
ISBN 0-8493-4377-1 (v.2)

Direct all inquiries to CRC Press, Inc., 2000 Corporate Blvd., N.W., Boca Raton, Florida, 33431.

© 1988 by CRC Press, Inc.

International Standard Book Number 0-8493-4375-5 (set)
International Standard Book Number 0-8493-4376-3 (vol. 1)
International Standard Book Number 0-8493-4377-1 (vol.2)

Library of Congress Card Number 87-28975
Printed in the United States

PREFACE

Herbicides constitute the major portion of sales for all pesticides in the highly organized agricultural production systems in the western industrialized world. Over 100 million ha of agricultural land is now being treated with herbicides each year in the U.S. alone. The behavior of herbicides, as opposed to pesticides in general, has been reviewed earlier under the editorships of Drs. P. C. Kearney and D. D. Kaufman of Beltsville, MD (1976) and by Dr. R. J. Hance of the Weed Research Organization, Oxford, England (1980), the former encompassing the environmental behavior in general, and the latter dealing with the soil component only. The rapid increase in the use of herbicides over the past decade, and the continuing introduction of new herbicides, as well as the expanding research efforts at universities, government institutions, and in the agrochemical industry to improve "efficacy" and "safety", necessitate an ongoing review process which summarizes information, and highlights gaps in information in order to keep abreast of this rapidly expanding field. This two-volume series should not only update available information on this subject, but also provide a better understanding of herbicide behavior in general, leading, hopefully, to increased efforts by researchers to fill in the remaining gaps in information.

Due to the vast data base being generated in this field, it was necessary to divide the subject matter into two volumes, and it was of course logical to compile all aspects of the soil component under one umbrella. Thus, Volume I deals with the physicochemical principles controlling the behavior of herbicides in the soil component of the environment, beginning with soil adsorption/desorption characteristics, diffusion and mass flow, and leading to transport as surface runoff and volatilization into the atmosphere. The last two chapters deal with the dissipation and transformations of herbicides in the soil. Each chapter deals with the physicochemical principles and processes controlling the behavior of herbicides, including a brief but critical account of the techniques associated with the subject matter at hand. This is followed by a summarizing of information for each herbicide class or group and tabulating this information wherever possible. The type and role of models being developed as predictive tools, and validation using field data whenever feasible, also receive special attention. Finally, each chapter concludes with a summary, which indicates gaps in information and includes suggestions for potential areas of future research.

The two-volume series is prepared for audiences such as weed scientists/agronomists involved in active research and/or making recommendations on weed control, environmental/residue chemists, engineers, and soil scientists associated with the study of herbicide behavior in the environment, agencies that are responsible for recommending and regulating the use of herbicides in agriculture and forestry, as well as those involved in their postapplication monitoring for environmental concerns.

FOREWORD

The use of herbicides has grown tremendously in the 40 years since the introduction of 2,4-D in 1945. Thousands of tonnes of several different phytotoxic compounds are now used annually around the world to control weeds. Their availability has contributed enormously to the increase in the quantity of food now produced globally and in the efficiency of food production. These advances have not only occurred in developed countries, but also in developing countries where it has been possible to raise the standards of living by reducing the endless hours of arduous labor previously required to control weeds by hand.

There is currently much interest in the development of integrated weed control systems wherein greater emphasis will be devoted to the incorporation of cultural and biological controls into recommended weed control practices. However, while this may reduce the farmers' dependence on chemicals, herbicides will undoubtedly form the backbone of weed control practices for the foreseeable future.

The benefits from this new technology are not without risks. There has to be concern over the introduction into the environment of such large quantities of compounds known to affect biological organisms. Indeed, this is well recognized and much of the efforts of weed scientists is now devoted to research on measuring and reducing the risks that are involved. Most countries now have rigid regulations requiring background information on the persistence and behavior of herbicides in the environment before new compounds can be registered for sale. The fate of older compounds is also being scrutinized. Fortunately, the new instruments that are available for these studies have made it possible to follow the adsorption, movement, and degradation of herbicides in extremely minute amounts. Thus, reliable data can now be obtained for the assessment of the impact of herbicides on the environment.

It is noteworthy that most of the published information referred to in this book has come from government or university laboratories. Unfortunately, a lot of the data developed by private companies, although available to regulatory agencies, does not get published. In spite of this, the authors have been able to present a good assessment of the state of our knowledge of the environmental chemistry of herbicides.

The information in these two volumes will be of interest to weed scientists and ecologists concerned with the impact of herbicides on the environment. It will be particularly useful to personnel responsible for the legislation concerning the registration of herbicides. And it will serve as useful reference material for teachers and students of weed science and toxicology.

Dr. R. Grover is to be commended for taking the initiative to asemble the information in this publication. He is well qualified to edit these two volumes. Since the early 1960s, he has been a leader in developing research on the fate of herbicides in the environment. He has published extensively on the movement of herbicides in soil and water, and on drift through the atmosphere. Most recently, he has been engaged in large scale mass balance studies following field applications of herbicides. He has been a leader in developing studies on the exposure of workers to herbicides. We are indebted to the publishers, Dr. Grover, and the authors whom he has selected for this timely addition to our knowledge in an important area of weed science.

J. R. Hay
Agriculture Canada
Research Station
Saskatoon, Saskatchewan

ACKNOWLEDGMENTS

I would like to thank Dr. James R. Hay, Director, Agriculture Canada Research Station, Saskatoon, Saskatchewan, and former President of the Weed Science Society of America, for encouragement to undertake this task and for taking time to write the Foreword for this volume. Advice of Drs. Raymond J. Hance, James R. Hay, Allan E. Smith, and William F. Spencer in selecting several of the authors is also acknowledged. Thanks are also due to Ms. Beverly Cleveland for valuable assistance in reading and editing the manuscripts and to her and Ms. Sherrie Racz for typing the manuscripts and ensuring that the publisher's guidelines were adhered to.

Raj Grover

THE EDITOR

Raj Grover, Ph.D., has been with the Research Branch, Agriculture Canada, since 1960, mostly at its Regina Research Station which is the national center for weed control research in Canada. For the past 16 years he has served as the Head of the Herbicide Behavior in the Environment Section at the Station. He has contributed to, or prepared briefs related to the environmental impact of herbicide use on behalf of several Canadian organizations, such as Agricultural Institute of Canada, Saskatchewan Institute of Agrologists, Expert Committee on Weeds, International Commission on Garrison Dam, etc. He is a member of several professional societies, such as Chemical Institute of Canada, Weed Science Society of America, American Chemical Society (Agrochemical Division), Society of Environmental Chemistry and Toxicology, American Society of Testing and Materials (E-35 Committee), and the Scientific Research Society of Sigma Xi. He has participated in special sessions or symposia organized by these and other groups on several occasions, again on topics related to herbicide behavior in the environment. He was also a visiting scientist at the Weed Research Organization, Oxford, England (1967/1968), and the Department of Soil and Environmental Sciences, University of California, Riverside (1976/77).

Raj Grover's main contributions to research concern various aspects of off-target transport of herbicides such as application drift, postapplication losses to air as vapor or in water as runoff or leaching to groundwater, air monitoring, etc. More recently he has been instrumental in initiating several multidisciplinary studies, such as Mass Balance of Herbicides Following Application, Exposure of Applicators to Herbicides, and Evaluation of New Application Technology. He has published over 60 scientific papers and has written or co-authored several chapters in monographs and books related to the environmental aspects of herbicide use.

CONTRIBUTORS

James M. Davidson, Ph.D.
Dean for Research
Institute of Food and Agricultural
  Sciences
University of Florida
Gainesville, Florida

Dwight E. Glotfelty, Ph.D.
Environmental Chemistry Laboratory
U.S. Department of Agriculture
Agricultural Research Service
Beltsville, Maryland

Raymond J. Hance, Ph.D.
Consultant
Brook Hill
Woodstock
Oxford, England

Ron E. Jessup, M.S.
Soil Science Department
Institute of Food and Agricultural
  Sciences
University of Florida
Gainesville, Florida

Ralph A. Leonard, Ph.D.
Southeast Watershed Research Laboratory
U.S. Department of Agriculture
Georgia Coastal Plain Experiment Station
Tifton, Georgia

Ralph G. Nash, Ph.D.
Pesticide Degradation Laboratory
U.S. Department of Agriculture
Agricultural Research Service
Beltsville, Maryland

P. S. C. Rao, Ph.D.
Soil Science Department
Institute of Food and Agricultural
  Sciences
University of Florida
Gainesville, Florida

Allan E. Smith, Ph.D.
Environmental Chemistry of Herbicides
  Section
Research Branch, Agriculture Canada
Research Station, Regina
Saskatchewan, Canada

Alan W. Taylor, Ph.D.
Agricultural Experimental Station
University of Maryland
College Park, Maryland

TABLE OF CONTENTS

Volume I

Chapter 1

ADSORPTION AND BIOAVAILABILITY

R. J. Hance

TABLE OF CONTENTS

I. INTRODUCTION

The ability of soil to remove solutes from solution is so well established that for generations adventurers in arid regions have known it is possible to "recycle" one's own urine for drinking by passing it through some sort of crude soil column. However, more precise studies have shown that the removal of solutes is not complete, but involves reversible processes which provide a wide range of ratios of distribution between the solid and liquid components. The equilibrium condition and the rate of its attainment are important factors affecting the biological availability of a chemical in the soil.

The purpose of this chapter is to outline the important general features of the physical chemistry involved and to show how they, together with the constraints imposed by processes controlling uptake by living tissues, would be expected to affect the interactions of herbicides in the soil with living organisms. This leads to a discussion both of situations where expectation is borne out in practice and of situations where it is not; which in turn generates some thoughts for possible future research.

II. ADSORPTION PROCESSES

Excellent reviews of the detail or processes by which organic molecules are adsorbed by soil are available,[1-5] so only a brief summary can be justified here, except where there have been significant recent developments.

A. Experimental Methods

A few words about experimental procedures are appropriate before beginning a theoretical discussion. There are four basic procedures that have been used for the assessment of the equilibrium distribution of a solute between adsorbent and solution. The most widely used is the "batch" method in which a slurry of soil and herbicide solution is shaken for a predetermined time, after which the soil is filtered off or centrifuged down and the herbicide concentration remaining in solution is measured. This method is specified in the guidelines for registering pesticides published by the U.S. Environmental Protection Agency.[6] It is simple and reasonably rapid (overnight shaking is usually long enough for equilibrium to be approached quite closely), but it suffers from two disadvantages. The first is that the adsorbed quantity is estimated by difference between the initial and equilibrium solution concentration, so, if adsorption is low, this difference will be small in relation to analytical error and the adsorption measurement will lack precision. The second is that the solution/ soil ratio in a slurry, typically between 5 and 20 to 1, is so different from the field condition that the results may be of questionable practical significance. In particular, slurry shaking causes soil particles to disperse so that a larger surface may be exposed than occurs in the field.[7]

The alternative methods all involve some sort of flow system. In two of them herbicide solution is allowed to percolate through a soil column. Green and Corey[8] allowed solution to flow through the column until the concentration of the effluent was the same as that of the applied solution. Then the occluded solution and adsorbed solute were displaced so that the amount adsorbed was measured directly. This method has good precision at low levels of adsorption and retains soil aggregate structure, but is time consuming since one column run produces only one data point. Burchill et al.[9] used a very short column, really a pad of soil, varied the input solute gradient, and monitored the effluent concentration continuously so as to allow calculation of adsorption at several concentrations in each run. As with the batch method the results will be imprecise at low levels of adsorption and, because concentrations change, there is uncertainty that equilibrium is achieved. The fourth of the methods was described by Grice et al.[10] and involves a passage of solution of known concentration through a cell containing adsorbent kept in suspension with a stirrer. Again, the effluent concentration is monitored and a series of measurements is possible from one suspension. This method allows an entire equilibrium adsorption and desorption isotherm to be produced from one experiment, but it requires special equipment and, like the batch slurry procedure, it disintegrates soil aggregates and gives poor precision if adsorption is low. It is, however, particularly useful for adsorbents of colloidal size which cannot be handled satisfactorily by the other methods.

B. Numerical Description of Adsorption and Desorption

The relationship between the amount adsorbed at constant temperature and the concentration of the solute in solution is usually called the adsorption isotherm. For a proper description, both solute and solvent should be considered, but, since in systems involving herbicides and soil the solute concentration expressed as a mole fraction is of the order of 10^{-4}, the effect of the solvent is usually assumed to be constant, although, as will be discussed later, water molecules play a significant part in adsorption processes.

There have been few attempts at rigorous theoretical treatments of herbicide adsorption by soil as most authors have been discouraged, not unreasonably, by the complexity introduced by the heterogeneity of the adsorbing surfaces in the soil. Consequently a preponderance of authors have used the empirical Freundlich isotherm:

$$\frac{x}{m} = k\, C^{\frac{1}{n}} \tag{1}$$

where x is the quantity adsorbed by mass m in equilibrium with concentration C; k and n are constants. Its only real merit is that at the low concentrations used in pesticide work, most sets of observations with nonionic molecules can be fitted as well as the accuracy of the data can justify. It would clearly be unsatisfactory at high concentrations since the adsorbed concentration (x/m) does not approach a limiting value as the equilibrium concentration (C) is increased without limit. Therefore, the relationship can be used only for *inter*polation not *extra*polation. A deficiency of more practical significance is that C is raised to an arbitrary fractional power so that the value of the constant k varies if n has a value other than 1. Thus, different adsorptions cannot be compared unless they have the same values of n. Fortunately for most herbicides the values lie between about 0.7 and 1,[1] so comparisons are usually possible for practical purposes, but there are exceptions, a notable one being adsorption by burnt straw ash.[11] This material has sometimes caused problems in minimum tillage cereal systems, particularly in the U.K., by increasing the adsorptive capacity of soil surface layers to such a level that some soil-applied herbicides become unreliable;[12] so clearly, a knowledge of its characteristics is of importance.

Other equations have from time to time been proposed,[1-5] but none has been adopted to any great extent except in the case of cationic materials, especially paraquat and diquat where ion-exchange equations must be used.[13]

Desorption isotherms are less easy to describe — even empirically. Desorption is usually a slower process than adsorption,[4] which can lead to significant errors caused by decomposition of the solute and modification of the adsorbent as a result of the length of the experimental period. Hence, it is frequently difficult to decide whether a particular set of observations demonstrates hysteresis, irreversibility, changes in the system, or differences in the rates of the forward and reverse processes. The subject is considered in more detail in the context of mass transfer in the next chapter.

Quantitative descriptions of adsorption in equilibrium systems have been invaluable in studies which have sought to understand the nature and extent of interactions between soils and solutes so that, as will be seen later, we can now classify to some extent the adsorption characteristics of both soils and herbicides. As a result, dosage rates can be prescribed with better precision. However, such prescriptions contain a large empirical content since it seems unlikely that equilibrium conditions are often approached following field applications of pesticides. Unfortunately, there is little direct evidence to substantiate this assertion, but observations of the behavior of radio-labeled ethirimol[14] and of herbicide concentrations in solutions expressed from soil in a pressure-membrane apparatus[15] indicate that soil structural effects may be more important than adsorption in influencing solute concentration in soil water.

C. Kinetics of Sorption

The effect of soil structure on adsorption is probably related to nonattainment of equilibrium, and, at least partly as a consequence, studies of adsorption kinetics do not yet allow an adequate quantitative description of events. Hamaker and Thompson[1] considered the theoretical aspects of sorption kinetics in some detail, and their discussion raised several experimental possibilities; few, if any, of which appear to have been pursued. In slurry systems adsorption by soil or soil components usually reaches equilibrium within 24 hr or less[2] with desorption rather slower. The practical significance of such observations is of doubtful significance since following a spray application, soil solution concentrations depend on the moisture content of the soil at the time of application as well as the degree of soil aggregation.[15,16] Also, there is evidence that desorption is slower (or the sorption is less reversible) following wetting and drying cycles of the sort that occur during weathering.[17]

At first sight there is a temptation to dismiss these uncertainties in the context of bioavailability since the time scales involved are often not very long compared with the periods over which soil-applied herbicides remain effective. This would probably be a mistake, because adsorption and desorption are important factors controlling the movement and distribution of chemicals in the soil (see Chapter 2), which in turn affect availability. Therefore, a better understanding of the kinetics and reversibility of sorption in structured field soil is needed for a proper assessment to be made of the importance of the basic process.

D. Mechanisms of Adsorption

The most important soil components for the adsorption of herbicides are the colloidal fractions, which are made up predominantly of clay minerals and organic matter. Descriptions of these materials can be found elsewhere.[1,5,18] The surfaces of pure clay minerals are negatively charged and exchange cations are present. They are also hydrophilic because of the solvation of the exchange ions and because of the high density of hydroxyl groups. They are composed of layered structures, and the layers of some minerals can separate so that the crystal expands and internal surfaces become accessible to molecules of a suitable size. Soil organic matter is essentially an aromatic polymer, more hydrophobic than clays, with the predominant functional groups carboxyl, hydroxyl, carbonyl, and alkoxyl. Both materials have the potential to adsorb molecules by ion exchange, coordination with metal exchange ions, hydrogen bonding, physical forces, and entropy effects. Some organic matter components can, in addition, act as surfactants. Many of our current ideas about how molecules are adsorbed are based on experiments done with pure clay minerals or isolated organic matter preparations, but it is important to remember that in the soil, humified materials and clay minerals are always in association. The nature of these associations is not well understood, but their adsorptive properties do not seem to show a simple relation to those of the individual constituents.[2,5]

Details of the possible interactions involved in adsorption can be found in the reviews mentioned previously,[1,3-5] or in Hayes[19] and Weber.[20] The bonds involved may conveniently be put into two energy categories with the division at about 80 kJ/mol. Covalent bonding would come into the high energy category, but perhaps is best considered as involving a transformation, rather than adsorption, and so is not discussed here.

1. High-Energy Bonds

a. Ionic Bonds

Since soil contains positively and negatively charged sites, it follows that ionic molecules can be adsorbed by ion exchange. Most herbicides are uncharged, but some are weak acids and others, such as the triazines, are weak bases.

Overall, the soil is negatively charged so adsorption of anions is relatively uncommon. However, iron oxides can carry a positive charge and hence can adsorb anions, such as 2,4-D, at pH values below the point of zero charge of the oxides.[21,22] Glyphosate, which is zwitterionic, can also be adsorbed as an anion,[23] although in this case, the interpretation is clouded by the fact that its iron and aluminium salts are of low water solubility.

Two common herbicides which are permanent cations are paraquat and diquat. These are adsorbed very strongly by clay minerals because their geometry allows the ionic bonding to be reinforced by short range forces such as charge transfer and Van der Waals forces.[24] This adsorption is so strong, and the equilibrium distribution is such, that for most practical purposes it can be regarded as irreversible, but in fact it is not, as Knight and Denny[25] showed complete self exchange occurred between ^{14}C-labeled paraquat in solution and unlabeled paraquat adsorbed on montmorillonite.

The other group of herbicides for which ion-exchange can be important is the 1,3,5-triazines, because they can accept protons and hence become cationic.[20] The pK_a for this reaction varies from 1.65 for simazine to over 4 for some of the methoxy and methylthio analogues, and maximum adsorption by clays occurs when the pH of the system is close to the pK_a. Although these values seem low compared with the pH of suspensions of normal soils, the surfaces of clays are much stronger proton donors than would be indicated by the pH of their aqueous suspensions,[26] so ion exchange could make a significant contribution to the adsorption of triazines in some practical situations.

b. Ligand Exchange

Ligand exchange onto partially chelated transition metals on clays and humic acids has been proposed as a possible binding mechanism for triazines and perhaps substituted ureas, aminotriazole, and EPTC.[2] The displacement of a weaker ligand, such as water, according to the following scheme is possible.[1]

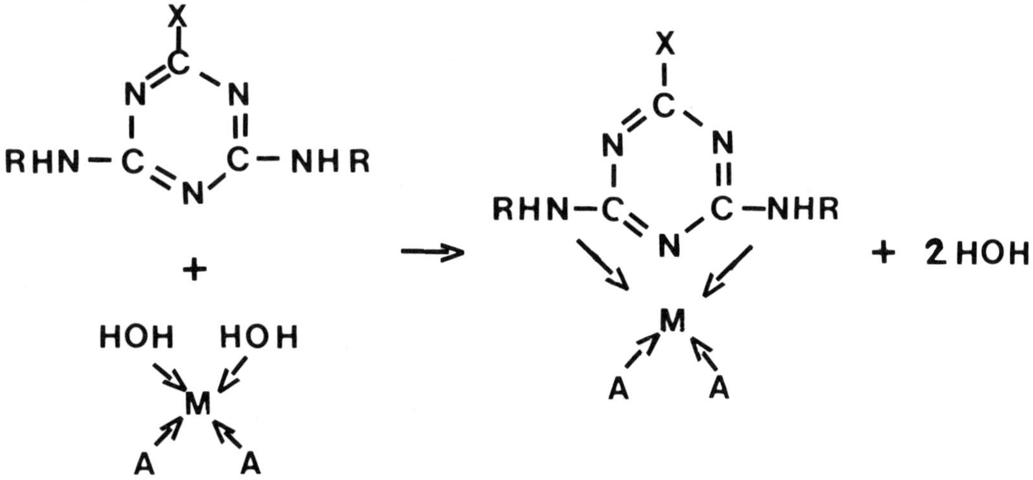

2. Low-Energy Bonds

a. Hydrogen Bonds

Hydrogen, covalently bonded to an electronegative atom, usually N or O, can be electrostatically attracted to another electronegative atom to form a bridge. Virtually all herbicide molecules contain groups that have the potential to form hydrogen bonds, but the role of such bonds in adsorption is uncertain. When water is present it is likely that a herbicide will form stronger bonds with it than with an adsorbent. However, the possibility of hydrogen-bonded water bridges between solute and adsorbent seems feasible, although at the moment, there is only indirect evidence for its occurrence.

b. Charge Transfer

Charge transfer complex formations occur when partial overlap of the molecular orbitals of an electron-rich and an electron-deficient molecule produces a partial exchange of electrons to form a resonance structure of ionic forms of donor and acceptor. Donor structures either contain *pi* bonds and electron rich substituents or lone pairs of electrons. Acceptors may be compounds with *pi* bonds and electron withdrawing groups or weakly acidic hydrogens. Thus, the hydrogen bond is a special case of charge transfer. There is spectroscopic evidence for charge transfer complexes between paraquat and diquat with montmorillonite and humic acid,[5] and it is likely that other compounds, particularly the triazines, behave similarly.

c. Charge-Dipole and Dipole-Dipole Bonds

Electric charges on adsorbents and exchange ions can attract polar molecules. Some large organic molecules are easily polarized; so, while such induced forces are usually small, it is possible that in some circumstances they could play a significant part. This, however, is purely speculation in the herbicide context.

d. London-Van der Waals Forces

These forces arise from fluctuations in electron distribution, producing transient dipoles which cause attractions. They are weak, being proportional to the sixth power of the inter-molecular distance, and by themselves are probably insignificant. However, they could reinforce attractions brought about by stronger forces.

E. Thermodynamic Considerations

The previous section outlines the major specific interactions that could bring about adsorption. However, they do not take account of the behavior of water which results from the powerful attraction between water molecules. Liquid water has properties that are reasonably well described by a model which involves a mixture of hydrogen-bonded clusters, possibly ice-like in structure, and a liquid which contains more densely packed, but less hydrogen-bonded molecules.[27] If a nonpolar molecule, such as a hydrocarbon, is introduced into water, heat is evolved, which implies that the entropy decrease is greater than the reduction in enthalpy. This suggests that water is restructured around the hydrocarbon to form an ice-like structure with more hydrogen bonds than occur in free water. If the hydrocarbon interacts with a nonpolar adsorbent, some of the structured water will be displaced and revert to its normal, less-structured state; hence, the entropy of the system is increased so that adsorption is thermodynamically favored. Thus, pesticides with nonpolar regions can be adsorbed onto hydrophobic components of soil organic matter as a result of weak attractions of the Van der Waals-type, reinforced by the entropy-generating effect. This process, which is often called "hydrophobic bonding" (clearly a misleading term), is probably of great importance in the adsorption of many herbicides, although only recently has direct experimental confirmation begun to appear.[28,29]

F. Soil Characteristics

The processes described above could, with the possible exception of anion exchange, occur on either clay or organic matter surfaces. For any particular herbicide, the relative importance of clays and organic matter depends on the balance between these processes and on the association between clay and organic components which will vary from soil to soil. For the bipyridinium compounds it is the clay fraction which is of predominant importance because the geometry of these molecules allows ion-exchange and entropy generation to be reinforced strongly by other forces. These compounds are also extensively adsorbed by organic matter, but the interactions are not so strong. In some soils, those in which montmorillonite is the predominant clay mineral, triazine adsorption is also dominated by the clay fraction.

With most nonionic herbicides the extent of adsorption is best correlated with soil organic matter content, although, as Calvet et al.[3] point out, this correlation is not always very good over the range 0 to 4% organic matter, which includes most temperate arable soils. It is not unreasonable to expect that organic matter should be important with these compounds, since they contain hydrophobic regions, so entropy generation could be important in their adsorption on hydrophobic soil organic matter components and, as Chassin et al.[30] have suggested, charge transfer and dipole, if not ionic, interactions are also likely.

The probable significance of specific interactions means that very good correlations cannot be expected between adsorption and soil organic matter content, because the interactions will vary depending on the nature of the compound and the source of the organic material. Nevertheless, the assumption that organic matter is largely responsible for the adsorption of organic molecules has been made by many workers who have followed the examples of Furmidge and Osgerby[31] and Lambert[32] and expressed adsorption distribution constants in terms of weight of organic matter rather than weight of soil. Soil-to-soil variability of this parameter is much lower than values calculated on a total soil basis, and, as will be seen in Section G, it has been particularly useful in deriving general relationships between adsorption and solute structure. It has also been used in agricultural practice in that some manufacturers' label recommendations for herbicides and the U.K. Agricultural Development and Advisory Service, vary dosage rate on the basis of soil organic matter content, or soil texture which is correlated with it.[33] The use of organic matter content, which is estimated from measurement of total carbon, can be misleading in soils containing straw ash, because charcoal residues are more adsorbent than soil organic matter.[12]

Anionic herbicides are generally not much adsorbed by soil, as would be expected, any adsorption usually being attributable to the hydrated metallic sesquoixides.

G. Herbicide Characteristics

This section is concerned only with uncharged herbicides which include virtually all the soil applied chemicals. Their adsorption may be expressed as a function of organic matter (k_{om}) from the relation:

$$K_{om} = \frac{\%100\ K}{\%\ OM} \tag{2}$$

where K preferably should be the Freundlich k (Equation 1), but is often the distribution ratio K obtained from an observation at only one concentration, using the assumption that the value of l/n in the Freundlich relationshp is close to 1. It has been known for a long time that adsorption was related to the lipophilicity of a molecule, and could be correlated with octanol/water partitions or reverse-phase t.l.c. measurements;[34] and Briggs[35] has shown that the Collander[36] relationship between partition coefficients can be applied:

$$\log K_1 = a \log K_2 + b \tag{3}$$

so that K_{om} can be treated as a soil organic matter-water distribution. This can be extended to show relationships between K_{om} and water solubility, and bioconcentration factors. Kenaga and Goring[37] came to similar conclusions on the basis of regression analysis of published data.

Despite the approximations and uncertainties involved, K_{om} for a chemical varies by only fourfold when considering soils world-wide, and prediction of one partition from another using the Collander equation, gives a value within an order of magnitude of observation.

Briggs[35] also went on to extend the idea that, since in order to disolve, a solute must make a hole in the solvent, the bigger the molecule, the lower the solubility. Thus, parachor, which is a measure of molecular volume, can be correlated with adsorption; if necessary, making corrections for hydrogen bonding or other solute-solvent interactions.[38] The parachor can be calculated from a knowledge of molecular structure so an initial assessment of the likely adsorptive behavior (and environmental partitioning) is possible with a paper exercise.

An alternative proposal for predicting soil sorption from molecular structure is the use of molecular topological indices.[39] Sabljic[40] calculated the first-order molecular indices ($^1\chi$) for 37 polycyclic aromatic or chlorinated hydrocarbons from the expression:

$$^1\chi = \sum_{s=1}^{n} (\delta_i \delta_j) s^{-0.5} \tag{4}$$

where δ for each nonhydrogen atom is the number of adjacent nonhydrogen atoms, and i and j correspond with the pairs of adjacent nonhydrogen atoms. The summation was carried out over all bonds between nonhydrogen atoms.

The correlation coefficient between log K_{om} and $^1\chi$ for the 37 compounds was 0.973. The $^1\chi$ index also correlated well (0.921) with the Van der Waals surface areas of the molecules. Correlation of log K_{om} with octanol-water partition was not so good, so it was inferred that adsorption of these molecules was a surface area-dependent process, rather than solute partitioning. The general applicability of this procedure remains to be established by studies with more polar and less planar molecules, but these results are undoubtedly promising.

Estimates of general partitioning behavior are very useful in the evaluation of environmental risk where prediction of detailed performance is not necessary. Together with an assessment of the stability of a molecule, which can be obtained from examination of its structure, they allow rough predictions to be made of persistence, mobility, and distribution in a range of conditions. As will emerge later, however, they often lack the precision needed to describe specific events.

III. UPTAKE PROCESSES

A. Plants

In common with many related subjects, studies of herbicide uptake by plants are bedeviled by the difficulty of designing experiments which do not disturb the system they are intended to study. In this case, there is the additional difficulty that the activity of the herbicide will modify the performance of a susceptible plant and hence may affect the uptake pattern. Thus, results obtained with nontoxic dosage rates are likely to be different to those obtained at biologically significant rates. Therefore, this restriction should be kept in mind throughout the following discussion.

1. Sites of Uptake

There seems no reason to suppose that any subterranean plant organ should not be capable of taking in an exogenous molecule, and there is evidence to support this view.[41] With a seedling or adult plant we would expect intuitively that the root system would be the most important site of entry by analogy with nutrient uptake and because the mass flow of water to the roots brings with it solutes. This certainly does occur, although it should be noted that not all herbicides enter roots at a rate proportional to water uptake.[42] In addition, as they grow, roots can encounter new supplies of herbicide that were not available to the germinating seedling. Therefore, in established plants, roots generally provide the major path of entry, although this generalization can be confounded by the relative distributions of herbicide and roots (see Section IV.), and also, if the herbicide has appreciable volatility, as under these circumstances, soil-applied materials may enter the plant through aerial parts.

2. Mechanism Uptake

Most soil-acting compounds seem to enter the plant passively and be transported in the apoplast (cell walls and xylem system). Therefore, it is possible to picture their transfer from soil to leaf as a sequence of partitions, starting with that between soil water and root, and continuing with those between transpiration stream and the various plant tissues.

The root-soil water partition has been designated the root concentration factor (RCF)[43] which is given by:

$$RCF = \frac{\text{Concentration in roots}}{\text{Concentration in external solution}}$$

As discussed earlier, partition ratios can be related by the Collander equation. The RCF increases with increasing lipophilicity, although for polar compounds, it does not fall below a limiting value of about 0.8. This has been explained by postulating that two processes occur.[44] One is partitioning between water and lipophilic root components, and the other, a simple equilibration between external water and the water in the cells and free space of the roots. Thus, a correction factor is needed in the Collander equation which, for a range of O-methyl carbamoyloximes and substituted phenyl ureas, gave the relation:

$$\log (RCF - 0.82) = 0.77 \log K_{ow} - 1.52$$

where K_{ow} is the octanol/water partition.

The efficiency of translocation is given by the transpiration stream concentration factor (TSCF),[43] defined as:

$$TSCF = \frac{\text{Concentration in the transpiration stream}}{\text{Concentration in external solution}}$$

Thus, for passive uptake, TSCF would have a maximum value of 1. Very polar and very lipophilic molecules are not well translocated, and maximum mobility seems to occur with compounds of K_{ow} about 1.8.[44] Even for those, TSCF rarely exceeds 0.8 because of the partitioning process in the root. Virtually all preemergence translocated herbicides have K_{ow} values between 1 and 3,[45] although compounds that are germination or shoot inhibitors, such as trifluralin or pendimethalin, may have higher values.

B. Uptake by Other Organisms

Organisms other than plants will take up herbicides from the soil water by eating food-containing residues or by topical application from spray drops or vapor. Much of the relevant material is covered elsewhere in the book, but a few comments may be useful in this chapter.

The distribution of a herbicide in the field is very uneven in relation to the size of microorganisms and small soil fauna. Thus, although the partition of a compound between soil solids, soil water, and living tissue should lead to similar accumulation levels in all soil organisms in any particular soil, this is unlikely to occur in practice because of the time needed for a uniform distribution of the herbicide to be attained. This assumes uptake is a passive process, as seems to be the case for earthworms[46] and, almost certainly, microorganisms. Also, on this assumption, because uptake into tissue would be proportional to K_{ow} and soil water concentration is inversely related to K_{ow}, all but the most polar compounds would accumulate to similar levels in living tissue in any particular soil and there would be an inverse relationship between soil organic matter content and concentration in tissues.

The situation with mobile organisms that are predaceous or graze selectively on organic materials is, of course, different. Not only does their mobility increase their chances of contacting high local herbicide concentrations, but their feeding habits give rise to the possibility of the food chain accumulation phenomenon.

IV. DISTRIBUTION AND BIOAVAILABILITY

Herbicide applied to the soil surface moves with the mass flow of water, following rainfall or irrigation, at a rate dependent upon adsorption and the flow of water (see Chapter 2). In temperate regions, redistribution throughout the plough layer by leaching is unlikely to occur with commonly used herbicides[45] and the maximum concentration will occur within a few centimeters of the soil surface. It is also worth noting that spring applications just precede a period of several months when evapotranspiration exceeds rainfall. Thus, although the root system is the most effective part of a plant for herbicide uptake, it may be exposed to lower concentrations than the shoot. If a crop is sown deeper than surface germinating weeds, there is, therefore, an opportunity to obtain some selectivity of herbicide action, often called "depth protection".

This simple view can, however, be misleading because as a plant grows, its roots may explore surface layers of the soil. For example, radish seedings sown 1 cm deep can produce roots spreading to 5 mm from the surface within 2 weeks;[47] and even perennial crops develop roots near the surface if they are not disturbed by cultivation.[48]

To a first approximation, therefore, it is possible to work on the assumption that uptake will be a function of concentration and the length of roots exposed.[49] The concept of concentration in this context needs to take account of the available soil water, proposed by Green and Obien,[50] to be that held between pH 2.5 to 4.2. It is likely that this assumption can be extended to other underground tissues, but they will make a significant contribution only when the root exposure $\times$ concentration product is low.

There are, however, examples where shoot uptake is of major importance. The classic example is that of tri-allate and wild oat *(Avena fatua)*, where the plant has a mesocotyl which raises the sensitive meristematic region into the tri-allate-treated surface soil at an early developmental stage, whereas wheat is tolerant because it has no mesocotyl so the meristematic region is protected by leaf-sheath bases.[51] Less obvious is the situation with many broad-leaved weed seedlings which are susceptible to photosynthesis inhibitors. The action of these compounds produces excited chlorophyll molecules and oxygen species that cause extensive tissue destruction. This occurs as soon as photosynthesis starts, so the plant may die within a few hours of emergence. At this stage the shoot may be of comparable length to the root so, presumably, is an important site of uptake.

The activity and selectivity of a herbicide may be modified by mechanical incorporation with the soil. For volatile or photolabile compounds this is, in any case, essential; but sometimes other compounds are also treated in this way so as to distribute them throughout the seed germination/rooting zone. This procedure reduces the possibility of depth protection, so can increase the spectrum of susceptible species. It may also be used as a means of reducing the activity of unwanted residues by dilution.[52]

A comment has already been made concerning mobile soil fauna, but it is necessary to recall that the distribution of both mobile and nonmobile species is neither even nor random. Most groups cluster to some extent in response to a variety of influences including egg-clumping, adverse conditions, local high concentrations of food or moisture, light, soil texture, and temperature. Because such distribution patterns change continuously, although they may show diurnal or seasonal periodicity, it is difficult to generalize about the distribution of herbicide and fauna with respect to exposure levels.

V. QUANTITATIVE ASPECTS

In order to make any sort of prediction of the environmental effects of a herbicide in any paricular situation, knowledge of its intrinsic biological activity is needed. Toxicology for insects and animals seems to be better developed than that for plants and microorganisms. This may be deceptive. The acute toxicity of a compound to an insect, for example, can be expressed in terms of mg/kg body weight following studies of the death rate of insects treated, topically or by injection, with a range of doses. This is more difficult with plants for a variety of reasons, including problems in defining when a plant is dead, great variations in response, depending on the site of application, and the concentration of the applied solution as well as the effect of environmental factors such as light, temperature, and humidity. For soil-applied materials, these uncertainties are increased by the effects of the various soil factors. The major problem, however, is to know how much material enters the plant. Even if this is known there may be difficulties over interpretation because, as discussed earlier, some fraction will partition into the plant tissue, and some of this will be adsorbed onto nonliving cell wall and xylem structures. Should such "inactivated" categories be counted as part of the dose or not?

For these reasons phytotoxicity assessments are most often recorded in terms of concentrations in the growth medium (or spray application rate per unit area of *land* rather than plant) which cause specified growth reductions compared with untreated control. Since the greatest change in response with change in dose rate occurs in the region of 50% response, it is common to use the quantity needed to produce a 50% response, designated variously as ED_{50}, GR_{50}, or ID_{50} as the criterion of toxicity, although for some purposes, ED_{10} or ED_{90} is more appropriate. Growth may be measured as shoot length, plant weight, or sometimes visual scoring systems are used. Discussions of the statistical treatments involved may be found elsewhere.[53,54]

This approach is used both in the evaluation of the properties of herbicides[54] and in bioassay procedures for measuring residues.[55] It is important to remember, however, the results obtained in uniform laboratory or greenhouse conditions will not be reproduced accurately in the field. Activity and selectivity usually decrease with the transition from a relatively controlled system to the field.[56]

The influence of soil, of course, further distorts the toxicity assessment, and for this reason activity is sometimes assessed in solution culture. A major drawback is that under these conditions, root habit and function are likely to be different from those which occur in soil. However, it is interesting that Hilton and Nomura[57] found that the minimum lethal concentration of most ureas and triazines to wheat, cucumber, and sorghum after 11 days exposure was of the order of 10^{-5} to 10^{-6} M, as this gives some idea of the quantities likely to be involved. Wetasinghe[58] found ED_{50} values of about 10^{-7} M of simazine and propazine against wheat, barley, and oat when plants were exposed for 72 hr, then transferred back to clean culture solution, figures that are satisfactorily close to those of Hilton and Nomura.

Although the measurement of TSCF (Section III.A.2.) requires the determination of quantities taken up, it is not relevant in this context because such studies involve sublethal concentrations in order to avoid the effects of herbicide toxicity on the physiology of the plant. Indeed there seems to be little literary information to relate quantity taken up (as opposed to quantity available to be taken up) with plant response. One recent example,[47] however, showed that less than 3 ng of chlortoluron reduced the weight of *Avena fatua* plants after 12 days by 17 to 40%, while over 30 ng per plant did not affect the growth of radish. More information of this sort would greatly improve our ability to assess the environmental significance of herbicides.

Toxicity to soil microorganisms is arguably even more difficult to assess. Measurement of total populations of soil organisms and studies in pure culture involve such artificial systems that they are of questionable relevance,[59] so current thinking suggests that measurement of processes such as nitrogen transformations and mineralization is more useful. This introduces the problem that naturally occurring stresses, such as drought and waterlogging, commonly reduce these activities by over 50%, but they ultimately recover. It has therefore been suggested[60] that, assuming a microbial population can double in about 10 days, recovery from a reduction to 1/8 of the original population would occur in 30 days (i.e., 3 doubling times) under favorable conditions, so effects lasting less than 30 days can be regarded as of "no ecological significance". Although very different from orthodox toxicological concepts, it is proving to be a useful approach to the assessment of herbicide effects on microorganisms.

Although simple acute toxicity assessments are possible with many components of the soil fauna, as with microorganisms, they are of dubious relevance in evaluating field responses. In addition to mortality, changes in behavior, growth, morphogenesis, feeding, reproduction, and longevity must be considered. Whether laboratory observations of these functions give valid answers are a cause for concern, and even if they do, integration of them to enable field predictions to be made is fraught with difficulty. This has not discouraged many workers from doing such work, as is plain from the review of Eijsackers and Van der Bund.[61] Direct field studies introduce additional methodological problems and further interactions with factors such as soil type, soil structure, tillage, fertilizers, cropping, and climate as well as the effect of other organisms. Many workers have been undeterred (Reference 61), but it is perhaps best to consider their output to be qualitative, rather than quantitative, and indeed it is difficult to imagine that anything else will be possible.

VI. CORRELATIONS BETWEEN ADSORPTION AND BIOAVAILABILITY

From the preceding sections it is clear that adsorption, per se, is only one of several factors affecting the availability of herbicides in the soil to weeds and nontarget organisms. It is of interest, therefore, to review briefly the situations in which adsorption seems to be the dominant influence and those in which it is not. In the case of plants, it is important to remember that where availability is inferred from some measurement of whole plant response (e.g., dry weight), a number of environmental factors, such as availability of water, temperature, humidity, and light, may be involved, as well as the factors discussed already. Thus, response may not be a very good indicator of availability.

A. Examples Where Correlations Occur
1. Plants

The most clear-cut examples of correlations between adsorption and availability are provided by the bipyridinium compounds, diquat and paraquat, where material adsorbed beyond the so-called "strong exchange capacity" of the soil is unavailable for all practical purposes,[62] although weaker adsorption by peat may leave some activity.[63] Other cases are reviewed by Summers.[13]

There are surprisingly few reported studies to compare availability of herbicides to plants with direct measurements of adsorption. In greenhouse experiments, Hance et al.[64] found reasonably good correlations between toxicity and adsorption for simazine with turnip and Italian ryegrass, lenacil and prometryne with ryegrass, and linuron with turnip, but not with lenacil and prometryne with turnip, or linuron with ryegrass. Walker[65] found that uptake of atrazine from 12 soils was proportional to soil water concentration calculated from slurry adsorption, although he later[66] observed with linuron that this did not always hold; a discrepancy he attributed to the failure of adsorption to come to equilibrium in his plant pots. Similar work with diuron,but with only two soils,[67] confirmed a relationship between solution concentration and uptake. Weber and Peter[68] obtained correlations between activity and adsorption for alachlor, acetochlor, and metalochlor, even though the adsorption isotherms were of different shape.

The most usual correlation to be made is that between soil organic matter content and activity, based on the assumption that soil organic matter provides the major sites for adsorption for most soil acting herbicides. A selection of these was tabulated by Walker[69] and showed correlations obtained with 11 compounds by nine authors in greenhouse experiments.

Even under controlled conditions caution is sometimes needed in studies of this sort since the plant itself may significantly perturb the system. As mentioned already, pH may affect adsorption and the plant may change the pH sufficiently during its growth to affect adsorption and hence, activity.[70]

Correlations in field work have usually been less convincing. In the classic multifactorial study of Upchurch et al.[71] involving five compounds applied for 3 years at 17 locations in the U.S., soil factors accounted for only some 60% of the total variation. Despite this, advisory and label recommendations frequently use either organic matter or soil textural class as the basis for prescribing application rates[33] with reasonable success. Interestingly, one circumstance in which organic matter gives a very poor guide to field performance is in minimum- or zero-tilled soils where crop residues are burnt, because the ash residue is highly adsorptive.[72] Measurement of carbon by wet oxidation includes not only organic matter, but also the more active elemental carbon in the ash, so the adsorption capacity of the surface soil is underestimated. Direct estimation of adsorption is therefore necessary.

Work in a different context is relevant here. It has been suggested[73,74] that the quantity of a herbicide that can be extracted from the soil with water should give a good indication of the availability of herbicide residues to plants. If this is so, then it should be possible to use such measurements to decide which crops can be safely grown in soil containing herbicide residues carried over from the previous crop, and so far, this approach has given promising results. Water extraction is basically a desorption procedure, so these studies provide some field evidence for a relationship between sorption and bioavailability.

2. Soil Microorganisms

Relations between adsorption and availability to microorganisms have mostly been studied with respect to rates of herbicide decomposition rather than toxic effects on microorganisms. Hence, most investigations provide only indirect evidence, because breakdown is not always exclusively microbiological, and a whole range of factors affect microbiological activity. One set of conflicting factors particularly clouds the issue. Soil microorganism activity is usually high in organic soils, but since adsorption is also generally related to organic matter levels, then availability of the herbicide and microbial activity will be negatively correlated. There is the additional consideration that the density of microorganisms tends to be higher near colloidal surfaces than elsewhere in the soil, so herbicide and microbe will be concentrated in the same region. Thus, prediction is difficult, though any sort of observation can be explained by recourse to an appropriate combination! Hurle and Walker[75] reviewed a range of conflicting results and concluded that, "adsorption does not always protect a chemical from degradation, nor does it always lead to increased rates of loss."

More analytical approaches to the study of breakdown are now appearing,[76,77] which provide frameworks which may allow the relative importance of the various factors, including adsorption, to be quantified.

There are three observations that perhaps are worth detailing here. One is that of Weber and Coble[78] who found that in nutrient solution diquat degradation was unchanged by adsorption on kaolinite, but when montmorillonite was present, the rate of breakdown was proportional to its concentration in solution. This is one of the clearest illustrations that there can be a distinction between *strength* and *extent* of binding. The second arises from experiments in which adsorption by soil was modified by the addition of either charcoal or montmorillonite.[79] Charcoal increased the adsorption of the three compounds studied, atrazine, chlorthiamid, and linuron. It reduced the rates of decomposition of atrazine and chlorthiamid, but not proportionally to the reduction in soil solution concentration, and it did not affect linuron breakdown. Montmorillonite increased the adsorption of atrazine and chlorthiamid, but not linuron, and slightly increased the persistence of all three. The explanation of results such as this with present knowledge requires a very fertile imagination! The third point of interest concerns the so-called "problem soils" which are soils where repeated applications of the same (or closely related) chemical has led to the development of microbial populations with an enhanced ability to degrade the compound so that the efficacy of the pesticide is reduced or lost. There are indications from studies with diphenamid[80] that this phenomenon is favored by low adsorption.

3. Other Organisms

Interactions between herbicides and organisms other than plants and soil microorganisms do not seem to have been reported in relation to adsorption in terrestrial soil. Chlorella has been used in a bioassay in adsorption studies,[81] which implies that the quantity of a herbicide which is in solution is available to it, and availability of the insecticide cypermethrin to *Daphnia, Cloeon,* and *Asellus* in water was reduced by adsorption by sediment.[82]

Studies of pesticide uptake by earthworms[46] led to the conclusion that availability should be related to adsorption by soil, but in practice, the calculated uptake is unlikely to be achieved because of the nonuniform distribution of the pesticide and metabolism in the worms.

B. Examples Where Correlations Apparently Do Not Occur

Since, in principle, adsorption controls soil solution concentration which in most circumstances should be related to bioavailability, a lack of correlation between adsorption and availability would be expected only if the system under examination does not attain equilibrium. For practical purposes, however, it is not availability, but biological activity that is of interest, and activity can be influenced by a number of processes other than adsorption. Hence, it is relevant here to consider examples where adsorption and activity are not related, that is, when other factors, which are usually climatic, override the influence of adsorption. Though there is extensive literature on the influence of the various environmental factors on herbicide performance, most of it describes experiments in which one factor is varied while the others are held constant. There is little work which varies two or more factors simultaneously, or directly attempts to assess the relative importance of each, including adsorption. The need to consider the effect of environmental and edaphic factors on persistence, which is an additional complication of particular importance in environmental considerations, has scarcely ever been met. These remarks are not intended to be overly critical; rather they are intended to show the difficulties involved. Blair[83] gives a more detailed analysis.

Although it is not possible to identify quantitatively specific factors which may override adsorption, it is nonetheless instructive to review a few examples where herbicide activity was clearly not entirely adsorption dependent. The work of Upchurch et al.[71] has already been mentioned and is perhaps the most comprehensive attempt to allocate relativities to the various factors. It was, however, a statistical approach, so does not necessarily allow causal deductions to be made, but it did suggest that soil organic matter was the major influence in that it accounted for some 60% of the variation in phytotoxicity observed. This has usually been interpreted to mean that adsorption was responsible, but it must be remembered that soil organic matter also affects microorganisms (hence, persistence) and soil structure (hence, root distribution and water movement) to mention but two properties that might affect plant response.

Hance et al.[64] made more direct attempts to relate adsorption to phytotoxicity by observing the toxicity of four herbicides in 16 soils to two indicator plants in field and greenhouse experiments. As discussed previously, even in the greenhouse they did not always obtain significant correlations between adsorption and phytotoxicity, and correlations with field phytotoxicity were even lower. It is also noteworthy that there were no correlations between ED_{50} values obtained in a spring field experiment with those obtained in the autumn on the same sites. Subsequent work[84] showed that toxicity on the same site could vary significantly with date of sowing and that even selectivity could be affected. Thus, there is clear evidence that adsorption is often not the variable which controls herbicide activity. This is tacitly admitted in manufacturer's label recommendations, where the variations in dosage rates suggested for different soil types are much smaller than variations in adsorption.[85] No doubt some of the potential effects of adsorption are offset by differences in plant growth, herbicide distribution, and exposure times. In addition, it seems likely that for many combinations of herbicide and weed, the normal dose rate is much higher than the minimum required for weed control.[69] The incorporation of such a margin for error into recommendations for farmers is prudent, but it is important to remember that when considering the effects on nontarget species, especially at residue levels, responses will be much more variable, and climatic factors are likely to be of greater significance.

VII. CONCLUSIONS AND FUTURE RESEARCH NEEDS

A high percentage of the work reviewed in this chapter was done for academic reasons, for example, that concerned with adsorption mechanisms, or with an agricultural purpose in mind. Applying this information to environmental issues may be difficult, but not necessarily impossible.

The area of most obvious relevance is that which seeks to relate adsorption to other partition functions, as this allows order-of-magnitude estimates to be made of the probable environmental distribution of a compound, a subject considered in detail elsewhere in the book. Thus, a broad estimate of bioavailability is possible. This approach is particularly valuable in preliminary assessment of the potential environmental hazards posed by a new compound.

Prediction or even assessment of the role played by adsorption in determining the bioavailability of a particular molecule in a specific situation is much more uncertain except, perhaps, in the growth chamber where all climatic variables can be controlled. Indeed, at the moment, perhaps all that can be said is that adsorption is certainly one of several factors which affect availability, the biological expression of which is affected by many more. Although theoretically it should be possible, using modern computer modeling techniques, to produce numerical descriptions and hence, predictions, the task of acquiring the data to use in this way seems to be beyond our current capabilities. A more realistic approach would be to generate enough data to enable general assessments to be made. Some of this information has already been accumulated, and this generalized approach is now in use with some success for the prediction of herbicide persistence.[75] It does, however, remain a daunting prospect requiring detailed monitoring of many interacting and fluctuating meterological and microclimatological factors in soils and plants.

Before such programs are established, a more basic question must be asked. Why do we need to know? In the narrow case of soil-acting herbicides we know enough, if only empirically, to use them reasonably effectively. The fact that we might be able to use smaller quantities if we understood better the effects of adsorption and other factors may be of economic significance, but the toxicology, persistence, and physical behavior of almost all herbicides suggests they will have little or no long-term environmental effects. The cultural practices that are possible because of the availability of herbicides may well have far reaching consequences, but a consideration of these must be sought elsewhere. Thus, although there are intellectual challenges associated with pesticide adsorption which, if overcome, may lead to refinement in use, more detailed study on environmental grounds may not be the most appropriate use of scarce research resources.

REFERENCES

1. **Hamaker, J. W. and Thompson, J. M.**, Adsorption, in *Organic Chemicals in the Soil Environment*, Vol. 1, Goring, C. A. I. and Hamaker, J. W., Eds., Dekker, New York, 1972, chap. 2.
2. **Calvet, R.**, Adsorption-desorption phenomena, in *Interactions between Herbicides and the Soil*, Hance, R. J., Ed., Academic Press, London, 1980, chap. 1.
3. **Calvet, R., Terce, M., and Arvieu, J. C.**, Adsorption des pesticides par les sols et leur Constituents, *Ann. Agron.*, 31, 125, 1980.
4. **Hartley, G. S. and Graham-Bryce, I. J.**, *Physical Principles of Pesticide Behaviour*, Vol. 1, Academic Press, London, 1980.
5. **Khan, S. U.**, *Pesticides in the Soil Environment*, Elsevier, Amsterdam, 1980.
6. Environmental Protection Agency, Guidelines for registering presticides in the United States. Chemodynamic parameters — adsorption, *Fed. Regist.*, 40 (123), 26881, 1975.
7. **Grover, R. and Hance, R. J.**, Effect of ratio of soil to water on adsorption of linuron and atrazine, *Soil Sci.*, 109, 136, 1970.
8. **Green, R. E. and Corey, J. C.**, Pesticides adsorption measurement by flow equilibrium and subsequent displacement, *Soil Sci. Soc. Am. Proc.*, 35, 561, 1971.
9. **Burchill, S., Cardew, M. H., Hayes, M. H. B., and Smedley, R. J.**, Continuous flow methods for studying adsorption of herbicides by soil dispersions and soil columns, in *Proc. Eur. Weed Res. Counc. Symp. Herbicides — Soil*, Versailles, 1973, 70.
10. **Grice, R. E., Hayes, M. H. B., Lundie, P. R., and Cardew, M. H.**, Continuous flow method for studying adsorption of organic chemicals by a humic acid preparation, *Chem. Ind. (London)*, p. 233, 1973.
11. **Embling, S. J., Cotterill, E. G., and Hance, R. J.**, Effect of heat treating soil and straw on the subsequent adsorption of chlortoluron and atrazine, *Weed Res.*, 23, 357, 1983.
12. **Hance, R. J. and Cotterill, E. G.**, Relationships between soil cultivation and pesticide performance, in *Monograph No. 27: Soils and Crop Protection Chemicals*, Hance, R. J., Ed., British Crop Protection Council, Croydon, 1984, 65.
13. **Summers, L. A.**, *The Bipyridinium Herbicides*, Academic Press, London, 1980, chap. 6.
14. **Collier, G. F., Graham-Bryce, I. J., Knight, B. A. G., and Coutts, J.**, Direct observations of the distribution of radio-labelled ethirimol in soil by resin impregnation and autoradiography, *Pestic. Sci.*, 10, 50, 1979.
15. **Hance, R. J. and Embling, S. J.**, Effect of soil water content at the time of application on herbicide content in soil solutions extracted with a pressure membrane apparatus, *Weed Res.*, 19, 201, 1979.
16. **Hance, R. J.**, The effect of aggregate size and water content on herbicide concentration in soil water, *Weed Res.*, 16, 317, 1976.
17. **Graham-Bryce, I. J.**, Adsorption of disulfoton by soil, *J. Sci. Food Agric.*, 18, 72, 1967.
18. **Greenland, D. J. and Hayes, M. H. B.**, *The Chemistry of Soil Constituents*, Wiley-Interscience, New York, 1978.
19. **Hayes, M. H. B.**, Adsorption of triazine herbicides on soil organic matter, including a short review on soil organic matter chemistry, *Residue Rev.*, 32, 131, 1970.
20. **Weber, J. B.**, Mechanisms of adsorption of *s*-triazines by clay colloids and factors affecting plant availability, *Residue Rev.*, 32, 93, 1970.
21. **Agustoni-Phan, N.**, *Adsorption of Herbicides on Iron Oxides*, Swiss Federal Institute of Technology, Zurich, Dissertation ETH 6266, 1978.

22. **Kavanagh, B. V., Posner, A. M., and Quirk, J. P.,** Effect of adsorption of phenoxyacetic acid herbicides on the surface charge of goethite, *J. Soil Sci.,* 31, 33, 1980.
23. **Hance, R. J.,** Adsorption of glyphosate by soils, *Pestic. Sci.,* 7, 363, 1976.
24. **Hayes, M. H. B., Pick, M. E., and Toms, B. A.,** Interactions between clay minerals and bipyridilium herbicides, *Residue Rev.,* 57, 1, 1975.
25. **Knight, B. A. G. and Denny, P. J.,** The interaction of paraquat with soil: adsorption by an expanding lattice clay mineral, *Weed Res.,* 10, 40, 1970.
26. **Mortland, M. M. and Meggitt, W. F.,** Interaction of ethyl N,N-dipropylthiolcarbamate (EPTC) with montmorillonite, *J. Agric. Food Chem.,* 14, 126, 1966.
27. **Franks, F.,** The role of water structure in disperse systems, *Chem. Ind. (London),* p. 560, 1968.
28. **Wauchope, R. D. and Koskinen, W. C.,** Adsorption-desorption equilibria of herbicides in soil: a thermodynamic perspective, *Weed Sci.,* 31, 504, 1983.
29. **Wauchope, R. D., Savage, K. E., and Koskinen, W. C.,** Adsorption-desorption equilibria of herbicides in soil: naphthalene as a model compound for entropy-enthalpy effects, *Weed Sci.,* 31, 744, 1983.
30. **Chassin, P., Calvet, R. and Terce, M.,** Adsorption de l'atrazine et du chlortoluron par les acides humiques, in *Proc. EWRS Symp. Theory and Practice of the Use of Soil Applied Herbicides,* Versailles, France, 1981, 10.
31. **Furmidge, C. G. L. and Osgerby, J. M.,** Persistence of herbicides in soil, *J. Sci. Food Agric.,* 18, 269, 1967.
32. **Lambert, S. M.,** Omega, a useful index of soil sorption equilibria, *J. Agric. Food Chem.,* 16, 340, 1968.
33. **Eagle, D. J.,** Matching herbicide dose to soil type, in *Pesticide Residues,* MAFF Reference Book 347, Her Majesty's Stationery Office, London, 1983, 86.
34. **Hance, R. J.,** Relationship between partition data and the adsorption of some herbicides by soils, *Nature (London),* 214, 630, 1967.
35. **Briggs, G. G.,** Theoretical and experimental relationships between soil adsorption, octanol-water partition coefficients, water solubilities, bioconcentration factors and the parachor, *J. Agric. Food Chem.,* 29, 1050, 1981.
36. **Collander, R.,** The distribution of organic compounds between iso-butanol and water, *Acta Chem. Scand.,* 4, 1085, 1950.
37. **Kenaga, E. E. and Goring, C. A. I.,** Relationship between water solubility, soil-sorption, octanol-water partitioning and bioconcentration of chemicals in biota, Third ASTM Symp. on Aquatic Toxicology, ASTM, 328, 1978.
38. **Hance, R. J.,** An empirical relationship between chemical structure and the sorption of some herbicides by soils, *J. Agric. Food Chem.,* 17, 667, 1969.
39. **Sabljic, A. and Trinajstic, N.,** Quantitative structure-activity relationships: the role of topological indices, *Acta Pharm. Jugosl.,* 31, 189, 1981.
40. **Sabljic, A.,** Predictions of the nature and strength of soil sorption of organic pollutants by molecular topology, *J. Agric. Food Chem.,* 32, 243, 1984.
41. **Schmidt, R. R. and Pestemer, W.,** Plant availability and uptake of herbicides from soil, in *Interactions between Herbicides and the Soil,* Hance, R. J., Ed., Academic Press, London, 1980, chap. 7.
42. **Sagar, G. R., Caseley, J. C., Kirkwood, R. C., and Parker, C.,** Herbicides in Plants, in *Weed Control Handbook: Principles,* 7th ed., Roberts, H. A., Ed., Blackwell Scientific, Oxford, 1982, chap. 3.
43. **Shone, M. G. T. and Wood, A. V.,** A comparison of the uptake and translocation of some organic herbicides and a systemic fungicide by barley. I. Adsorption in relation to physico-chemical properties, *J. Exp. Bot.,* 25, 390, 1974.
44. **Briggs, G. G., Bromilow, R. H., and Evans, A. A.,** Relationships between lipophilicity and root uptake and translocation of non-ionised chemicals by barley, *Pestic. Sci.,* 13, 495, 1982.
45. **Briggs, G. G.,** Factors affecting the uptake of soil-applied chemicals by plants and other organisms; in *Monograph No. 27; Soils and Crop Protection Chemicals,* Hance, R. J., Ed., British Crop Protection Council, Croydon, 1984, 35.
46. **Lord, K. A., Briggs, G. G., Neale, M. C., and Manlove, R.,** Uptake of pesticides from water and soil by earthworms, *Pestic. Sci.,* 11, 401, 1980.
47. **Addala, M. S. A., Hance, R. J., and Drennan, D. S. H.,** Studies on the site of uptake by plants of chlortoluron and terbutryne, *Weed Res.,* 25, 151, 1985.
48. **Atkinson, D. and White, G. C.,** Soil management with herbicides — the response of soils and plants, *Proc. 1976 Br. Crop Protection Conf. — Weeds,* 3, 873, 1976.
49. **Walker, A.,** Vertical distribution of herbicides in soil and their availability to plants: treatment of different proportions of the total root system, *Weed Res.,* 13, 416, 1973.
50. **Green, R. E and Obien, S. R.,** Herbicide equilibrium in soils in relation to soil water content, *Weed Sci.,* 17, 514, 1969.

51. **Walker, A., Briggs, G. G., Greaves, M. P., Hance, R. J., and Thompson, A. R.,** Herbicides in soil, in *Weed Control Handbook: Principles,* 7th ed., Roberts, H. A., Ed., Blackwell Scientific, Oxford, 1982, chap. 4.
52. **Eagle, D. J.,** Interpretation of soil analyses for herbicide residues, *Proc. 1978 Br. Crop Protection Conf. — Weeds,* 2, 535, 1978.
53. **Holly, K., Evans, S. A., and Rosher, P. H.,** The evaluation of a new herbicide, in *Weed Control Handbook: Principles,* 7th ed., Roberts, H. A., Ed., Blackwell Scientific, Oxford, 1982, chap. 7.
54. **Hurle, K. and Kemmer, A., Eds.,** *Biologische Testverfahren in der Herbologischen Forschung,* Heft 24, Universität Hohenheim, 1983, 163.
55. **Hance, R. J. and McKone, C. E.,** The determination of herbicides in *Herbicides; Physiology, Biochemistry, Ecology, Vol 2, 2nd Ed.,* Audus, L. J., Ed., Academic Press, London, 1976, 393.
56. **Riley, D. and Morrod, R. S.,** Relative importance of factors influencing the activity of herbicides in soil, *Proc. 1976 Br. Crop Protection Conf. — Weeds,* 3, 971, 1976.
57. **Hilton, H. W. and Nomura, N.,** Phytotoxicity of herbicides as measured by root absorption, *Weed Res.,* 4, 216, 1964.
58. **Wetasinghe D. T.,** *Studies on the Relative Toxicities to Plants of Some Triazine Herbicides,* Ph.D. thesis, University of Reading, Reading, Pennsylvania, 1966.
59. **Greaves, M. P. and Malkomes, H. P.,** Effects on soil microflora, in *Interactions Between Herbicides and the Soil,* Hance, R. J., Ed., Academic Press, London, 1980, chap. 9.
60. **Domsch, K. H.,** Interpretation and evaluation of data, in *Recommended Tests for Assessing the Side-Effects of Pesticides on the Soil Microflora,* Greaves, M. P., Poole, N. J., Domsch, K. H., Jagnow, G., and Verstraete, W., Eds., Technical Report Agricultural Research Council, Weed Research Organization (59), 1980, 6.
61. **Eijsackers, H. and Van de Bund, C. F.,** Effects on soil fauna, in *Interactions Between Herbicides and the Soil,* Hance, R. J., Ed., Academic Press, London, 1980, chap. 10.
62. **Riley, D., Wilkinson, W., and Tucker, B. V.,** Biological unavailability of bound paraquat residues in soil, in *Bound and Conjugated Pesticide Residues,* Kaufman, D. D., Still, G. G., Paulson, G. D., and Bandal, S. K., Eds., Symp. Ser., Vol. 29, American Chemical Society, Washington, D.C., 1976, 301.
63. **Damanakis, K., Drennan, D. S. H., Fryer, J. D., and Holly, K.,** Availability to plants of paraquat adsorbed on soil or sprayed on vegetation, *Weed Res.,* 10, 305, 1970.
64. **Hance, R. J., Hocombe, S. D., and Holroyd, J.,** The phytotoxicity of some herbicides in field and pot experiments in relation to soil properties, *Weed Res.,* 8, 136, 1968.
65. **Walker, A.,** Availability of atrazine to plants in different soils, *Pestic. Sci.,* 3, 139, 1972.
66. **Walker, A.,** Availability of linuron to plants in different soils, *Pestic. Sci.,* 4, 665, 1973.
67. **Moyer, J. R., McKercher, R. B., and Hance, R. J.,** Influence of adsorption on the uptake of diuron by barley plants, *Can. J. Plant Sci.,* 52, 668, 1972.
68. **Weber, J. B. and Peter, C. J.,** Adsorption, bioactivity and evaluation of soil tests for alachlor, acetochlor and metolachlor, *Weed Sci.,* 30, 14, 1982.
69. **Walker, A.,** Activity and selectivity in the field, in *Interactions Between Herbicides and the Soil,* Hance, R. J., Ed., Academic Press, London, 1980, chap. 8.
70. **Harris, G. R. and Hurle, K.,** The effect of plant-induced changes of pH upon the adsorption and phytotoxicity of *s*-triazine herbicides, *Weed Res.,* 19, 343, 1979.
71. **Upchurch, R. P., Selman, F. L., Mason, D. D., and Kamprath, E. J.,** The correlation of herbicidal activity with soil and climatic factors, *Weeds,* 14, 42, 1966.
72. **Cussans, G. W., Moss, S. R., Hance, R. J., Embling, S. J., Caverly, D., Marks, T. G., and Palmer, J. J.,** The effect of tillage method and soil factors on the performance of chlortoluron and isoproturon, *Proc. 1982 Br. Crop Protection Conf. — Weeds,* 1, 153, 1982.
73. **Stalder, L. and Pestemer, W.,** Availability to plants of herbicide residues in soil. I. A rapid method for estimating potentially available residues of herbicides, *Weed Res.,* 20, 341, 1980.
74. **Pestemer, W., Stalder, L., and Eckert, B.,** Availability to plants of herbicide residues in soil. II. Data for use in vegetable crop rotations, *Weed Res.,* 20, 349, 1980.
75. **Hurle, K. and Walker, A.,** Persistence and its prediction, in *Interactions between Herbicides and the Soil,* Hance, R. J., Ed., Academic Press, London, 1980, chap. 4.
76. **Frehse, H. and Anderson, J. P. E.,** Pesticide residues in soil-problems between concept and concern, in *Pesticide Chemistry: Human Welfare and the Environment,* Vol. 4., Myamoto, J. and Kearney, P. C., Eds., Pergamon Press, Oxford, 1983, 23.
77. **Soulas, G.,** Mathematical model for microbial degradation of pesticides in the soil, *Soil Biol. Biochem.,* 14, 107, 1982.
78. **Weber, J. B. and Coble, H. D.,** Microbial decomposition of diquat adsorbed on montmorillonite and kaolinite clays, *J. Agric. Food Chem.,* 16, 475, 1968.
79. **Moyer, J. R., Hance, R. J., and McKone, C. E.,** The effect of adsorbents on the rates of degradation of herbicides incubated with soil, *Soil Biol. Biochem.,* 4, 307, 1972.

80. **Kaufman, D. D., Katan, Y., Edwards, D. F., and Jordan, E. G.,** Microbial adaptation and metabolism of pesticides, in *Agricultural Chemicals of the Future, BARC Symp. No. 8.,* Hilton, J. L., Ed., Rowan and Allanhold, Totowa, New Jersey, 1983, 35.
81. **Lefebvre-Drouet, E. and Calvet, R.,** Utilisation des chlorelles pour le dosage biologique de l'atrazine dans le sol, *Weed Res.,* 22, 257, 1982.
82. **Riley, D. and Hill, I. R.,** Adsorption reduces activity of pesticides in soil and water, *Proc. 10th Int. Congr. Plant Protection,* Vol. 2, British Crop Protection Council, Croydon, 1983, 728.
83. **Blair, A. M.,** Some problems associated with studying effects of climate on the performance of soil-acting herbicides, in *Aspects of Applied Biology, Part IV, Influence of Environmental Factors on Herbicide Performance and Crop and Weed Biology,* Association of Applied Biologists, Wellesbourne, Warwick, 1983, 379.
84. **Hance, R. J.,** Processes in soil which control the availability of pesticides, *Proc. 10th Int. Congr. Plant Protection,* Vol. 2, British Crop Protection Council, Croydon, 1983, 537.
85. **Gerber, H. R., Nyffeler, A., and Green, D. H.,** The influence of rainfall, temperature, humidity and light on soil- and foliage-applied herbicides, in *Aspects of Applied Biology, Part IV, Influence of Environmental Factors on Herbicide Performance and Crop and Weed Biology,* Association of Applied Biologists, Wellesbourne, Warwick, 1983, 1.

Chapter 2

MASS FLOW AND DISPERSION

P. S. C. Rao, Ron E. Jessup, and James M. Davidson

TABLE OF CONTENTS

I. INTRODUCTION

The mobility of herbicides in soils directly determines their effectiveness as preemergence weed control agents as well as their potential for groundwater and surface water contamination. Herbicide mobility and persistence in soils have been measured by a number of scientists under laboratory and field conditions. These studies have laid the necessary foundations for detailed qualitative assessment of processes influencing herbicide behavior in soils. It is not our intent to review here this vast body of experimental data, especially since several recent articles and books have dealt with that topic.[1,2] Instead, we focus our attention on the quantitative aspects of the physical-chemical processes governing herbicide movement in soils. The application of these theoretical approaches for developing simplified models and indices useful in providing guidance for agronomic and environmental applications will be discussed. Finally, we will attempt to identify research directions that might lead to a better field-scale understanding of herbicide behavior in soils.

II. PRINCIPLES GOVERNING HERBICIDE TRANSPORT IN SOILS

A. Coupled Water and Solute Transport

Water is the main agent responsible for transporting nonvolatile herbicides and other agrochemicals such as fertilizers and salts in soils. (Vapor-phase movement of volatile herbicides is covered elsewhere in this book.) A description of water movement in soils is therefore a prerequisite to understanding herbicide transport. The fundamental physical principles governing water and solute movement in soils are well established and have been experimentally tested by numerous scientists over the past 2 decades. The historical perspectives of these developments and detailed aspects of the basic theories can be found in several review chapters and text books on the topic.[3-6] For this reason, we present here only a brief overview of the fundamental equations that have been used to describe water and solute flow in soils. This will define the parameters needed for a quantitative description of the processes, as well as provide a basis for a discussion of the limitations encountered in using the theory for field-scale problems.

Combining Darcy's law $[q = -K(h) \nabla \phi]$ with the continuity equation $[\partial\theta/\partial t = -\nabla q]$ results in the differential equation, known as the Richard's equation, for describing water flow:

$$(\partial\theta/\partial t) = \nabla.[K(h)\nabla\phi] \tag{1}$$

where q is the Darcian soil-water flux $(L^3 L^{-2} T^{-1})$; $K(h)$ is the soil hydraulic conductivity (LT^{-1}) as a function of h, the soil-water matrix potential (L); $\phi(=h + z)$ is the total soil-water potential (L); z is the vertical distance (L); t is time (T); θ is the volumetric soil-water content $(L^3 L^{-3})$; and ∇ is the standard differential operator.

A consideration of a fully three-dimensional flow is necessary for many practical applications, such as in herbicide application via drip irrigation.[2] The basic principles of water flow in soils can be understood equally well, however, by considering flow in only one direction. For vertical downward flow, the most common field situation, Equation 1 simplifies to:

$$\frac{\partial\theta}{\partial t} = \frac{\partial}{\partial z}\left[K(h)\frac{\partial\phi}{\partial z}\right] \tag{2}$$

A number of analytical and numerical solutions to Equations 1 and 2 for a variety of initial and boundary conditions are available.[4,5,7,8] More recently, based on a numerical solution to Equation 2, Nofziger[9] developed an interactive software for microcomputers. This software provides graphical outputs and allows the user to examine the impact of various initial and boundary conditions as well as soil hydraulic properties on water retention and flow in soils. The solutions to Equations 1 and 2 have been tested with data from laboratory-packed soil columns and field studies.[10-13]

Equation 2 needs to be modified to account for water uptake by plant roots. This is done by adding a "sink" term as shown below:

$$\frac{\partial \theta}{\partial t} = \frac{\partial}{\partial z}\left[K(h)\frac{\partial \phi}{\partial z}\right] - U(z,t) \tag{3}$$

where $U(z,t)$ represents the volumetric root uptake rate ($ML^{-3}T^{-1}$) and is a complex function of the crop root density, potential evapotranspiration rate, which depends on meterological variables, and soil hydrologic properties. Hillel[8] discusses various approaches for modeling water uptake by plants.

Movement of water through soils results in *mass flow* of water-soluble chemicals such as herbicides. The resulting herbicide flux (J_m) is simply the product of the Darcian soil-water flux (q) and the herbicide concentration (C; ML^{-3}) in soil-water. Thus,

$$J_m = qC = v\theta C \tag{4}$$

where J_m represents the amount of herbicide moving across a unit cross-sectional area of soil per unit time ($ML^{-2}T^{-1}$); v ($= q/\theta$) is the average pore-water velocity (LT^{-1}) and the other terms are as defined earlier.

Equation 4 represents only the *average* rate of herbicide flow in soils and does not account for the wide range in pore-water velocities encountered within and between pore domains that exist in soils as a result of variations in sizes and shapes of pores. Furthermore, herbicide concentration gradients exist between and within pore sequences. These factors contribute to another component of herbicide transport in soils — the *hydrodynamic dispersion*. At least three separate physical processes may contribute to hydrodynamic dispersion:

1. Molecular diffusion resulting from herbicide concentration gradients within a given pore sequence.
2. Mechanical dispersion (or mixing) resulting from variations in pore velocities among pore sequences.
3. Mixing resulting from diffusive transfer of herbicide between pore sequences having different velocities.

As will be discussed shortly, the relative effects of each of these processes is dependent upon pore geometry and specific flow conditions.

It is generally assumed that the effects of each of these processes are additive and that they can be described by Fick's first law. With these assumptions, herbicide flux due to hydrodynamic dispersion is usually described by:

$$J_{Dh} = -D_h\theta(\partial C/\partial z) \tag{5}$$

where

$$D_h = [D_e + D_m + D_s] \tag{6}$$

J_{Dh} is the herbicide flux ($ML^{-2}T^{-1}$) resulting from hydrodynamic dispersion; D_h is the hydrodynamic dispersion coefficient (L^2T^{-1}); D_e is the molecular diffusion coefficient (L^2T^{-1}); D_m is the "mechanical" dispersion coefficient (L^2T^{-1}); and D_s is "sink" diffusion coefficient (L^2T^{-1}). Note that the three terms on the right hand side of Equation 6 account for the three processes contributing to dispersion, as discussed above.

Davidson et al.,[14] among several others, have stated that,

$$D_e = D_o\alpha \tag{7}$$

$$D_m = \epsilon v \tag{8}$$

where D_o is the molecular diffusion coefficient in bulk water (L^2T^{-1}); α is the tortuosity or complexity factor (dimensionless); ϵ is the dispersivity factor (L); and other terms were defined earlier.

The coefficient α in Equation 7 accounts for the impedance to molecular diffusion within pore sequences resulting from the geometric complexity of soils. The numerical values for α range between 0 and 1; the smaller the α value, the greater is the impedence to diffusion in the pores compared to bulk water. D_e values for pesticides have been reported to be in the range of 10^{-2} to 10^{-1} cm² day^{-1}.[15-18]

The dispersivity factor (ϵ) in Equation 8 is an empirical index of the extent of variations in the pore velocities from the average velocity (v). Soils with narrow pore size and pore velocity distributions (such as sandy soils) are characterized by smaller values. On the other hand, finer-textured loamy and sandy soils with a wide range in velocities have larger ϵ values. Thus, the coefficient ϵ reflects our ignorance of the complexity of the soil pore geometry and the inability to specify a pore velocity distribution. Most laboratory column studies suggest that ϵ values are about a few millimeters ($\leqslant 1$ cm), whereas analysis of field data indicate that ϵ values may be in the order of several centimeters ($\leqslant 10$ cm).[19] The reasons for such variations in ϵ as a function of the scale of measurement are discussed further in a later section.

The term D_s in Equation 6 accounts for solute diffusion between pore domains having different velocities. D_s value depends on a number of soil and transport parameters, and as such, it is difficult to assign a specific value to it. However, these effects have been examined experimentally and theoretically and for certain special cases, analytical expressions for D_s may be derived. For example, in soils with uniform spherical aggregates, Passioura and Rose[20] and Rao et al.[21] have shown that the D_s term is directly proportional to v^2 and a^2, where a is the aggregate radius (L). Thus, with increasing aggregate radius (a), the contribution of "sink" effects due to solute diffusion into and out of the intra-aggregate regions becomes a more dominant factor. Rao et al.,[22] Addiscott,[23] and van Genuchten[24] have recently extended this analysis to consider other aggregate shapes and a distribution of aggregate sizes.

From the foregoing discussion it should be evident that as the pore-water velocity decreases (i.e., $v \rightarrow 0$), the major process of interest is molecular diffusion (D_e); the terms D_m and D_s in Equation 6 become negligibly small. The converse is true with increasing v and the diffusion term D_s may be ignored. Passioura and Rose,[20] Rao et al.,[21] and van Genuchten[24] have shown that in aggregated soils, the "sink" diffusion term (D_s) may be the most dominant compared with the contributions of molecular diffusion and mechanical dispersion.

The *total* herbicide flux, denoted by J_s $(ML^{-2}T^{-1})$, is the sum of fluxes resulting from mass flow (or convection) and hydrodynamic dispersion (or for brevity, dispersion). Thus, from Equations 4 and 5,

$$J_s = (J_{Dh} + J_m) = [-D_h\theta(\partial C/\partial z) + qC] \tag{9}$$

The conservation of mass principle for solute flow states that,

$$(\partial M/\partial t) = -(\partial J_s/\partial z) + \Omega(z,t) \tag{10}$$

where M is the total amount of herbicide per unit volume of soil (ML^{-3}); and $\Omega(z,t)$ represents herbicide losses or gains due to such processes as biotic/abiotic transformations, solubilization/precipitation, and plant uptake. Combining Equations 9 and 10 yields,

$$(\partial M/\partial t) = \frac{\partial}{\partial z} [D_h\theta(\partial C/\partial z) - qC] + \Omega(z,t) \tag{11}$$

For nonvolatile herbicides, the total mass, M, is given by the sum of the amounts present in the solution-phase and the adsorbed-phase. Therefore,

$$M = [\theta C + \rho S] \tag{12}$$

where ρ is the soil bulk density (ML^{-3}); S is the sorbed-phase concentration (MM^{-1}); and C and θ are as defined before. Substituting Equation 12 into Equation 11, we obtain,

$$\frac{\partial}{\partial t} (\theta C + \rho S) = \frac{\partial}{\partial z} [D_h\theta(\partial C/\partial z) - qC] + \Omega(z,t) \tag{13}$$

The above equation is valid for herbicide transport during vertical, transient water flow in soils. Solving Equation 13 requires specific knowledge of the changes in soil-water content (θ) and soil-water flux (q) as a function of soil depth (z) and time (t). These are obtained by first solving the Richard's equation for water flow (Equation 2), and are then used to solve Equation 13 to compute changes in herbicide concentrations (C and S) and fluxes (J_s) with time (t) and depth (z) in a soil profile. Thus, the coupling between water and herbicide flow is explicitly accounted for.

During steady water flow conditions, the soil-water content does not change with time $(\partial \theta/\partial t = 0)$ and the soil-water flux is the same at all depths $(\partial q/\partial z = 0)$. For such a case, Equation 13 simplifies to:

$$\theta \frac{\partial C}{\partial t} + \rho \frac{\partial S}{\partial t} = \theta D_h \frac{\partial^2 C}{\partial z^2} - q \frac{\partial C}{\partial z} + \Omega(z,t) \tag{14}$$

Solving Equations 13 or 14 necessitates that we specify the functional relationship between S and C, and expressions for Ω. The dependence of S on C at equilibrium is given by the sorption isotherm, $S = f(C)$, whereas a number of rate expressions may be used to describe the time dependence of sorption.[25,26] If we assume that the rate of sorption is rapid enough that essentially instantaneous equilibrium is achieved between the solution- and the sorbed-phases, and that a linear, reversible isotherm is appropriate, f(C) is given by,

$$S = K_D C \tag{15}$$

where K_D is the equilibrium sorption or partition coefficient (L^3M^{-1}). A number of authors have noted that for nonionic herbicides, the K_D value is determined primarily by soil organic carbon content (OC). Based on this observation, for each herbicide an OC-normalized sorption coefficient, denoted as K_{oc}, $(= K_D/OC)$ may be calculated. Literature compilations of K_D and K_{oc} values for a number of herbicides and related toxic organic compounds have been published[27-30] (see Chapter 1 for a more detailed discussion of herbicide sorption).

Herbicide transformations have most often been described by simple first-order kinetics, leading to the following expression for the term Ω:

$$\Omega = -[k_c \theta C + k_s \rho S] \tag{16}$$

where k_c and k_s are the first-order rate coefficients (T^{-1}) for herbicide degradation in solution- and sorbed-phases, respectively. Substituting Equation 15 in Equation 16 yields,

$$\Omega = -[(k_c \theta + k_s \rho K_D)C] \tag{17}$$

It is usually assumed that, $k_c = k_s = k$, leading to:

$$\Omega = -[k(\theta + \rho K_D)C] \tag{18}$$

In Equation 18 then, k represents the pooled rate coefficient for herbicide degradation via all pathways and in all phases. Literature compilations of such k values are available.[28] The importance of soil and environmental factors on k has been explored by Walker and associates.[31-39]

Now, we return to Equation 14, and substitute Equations 15 and 18 to obtain:

$$(\theta + \rho K_D)\frac{\partial C}{\partial t} = \theta D_h \frac{\partial^2 C}{\partial z^2} - q \frac{\partial C}{\partial z} - [k(\theta + \rho K_D)C] \tag{19}$$

Let us define:

$$R = [1 + (\rho K_D/\theta)] \tag{20}$$

$$v = (q/\theta) \tag{21}$$

$$D_h^* = (D_h/R) \quad \text{and} \quad v^* = (v/R) \tag{22}$$

Using these variables in Equation 19,

$$\frac{\partial C}{\partial t} = D_h^* \frac{\partial^2 C}{\partial z^2} - v^* \frac{\partial C}{\partial z} - kC \tag{23}$$

The above equation describes nonvolatile herbicide transport, sorption, and degradation in soils during one-dimensional, steady, vertical flow of water in homogeneous (not layered) soil profiles. Equation 13 is used to describe herbicide transport under transient water flow conditions. These two equations (or other forms) are commonly referred to as the convection-dispersion equations for solute flow.

Note that on the right hand side of Equation 23, the effects of hydrodynamic dispersion are accounted for in the first term, the mass flow effects are incorporated in the second term, while degradation losses are represented in the last term. van Genuchten and Alves[40] presented a compilation of analytical solutions to Equation 23 for a variety of initial and boundary conditions.

The parameter R in Equations 20 and 22 is known as the retardation factor and is an index for the effects of sorption on herbicide transport by mass flow and dispersion. Note that R values are directly proportional to the sorption coefficient (K_D). Therefore, with decreasing sorption $(K_D \rightarrow 0)$ R attains the limiting value of 1, suggesting no retardation.

On the other hand, as the K_D value increases, indicating increasing sorption, R value becomes very large $(R \gg 1)$.

At this point, it is useful to recall that v^* represents the average rate at which a herbicide moves through the soil as a result of mass flow. Thus, the time needed to travel some distance L by convective flow is given by $t_v = (L/v^*)$. Similarly, the time required to travel the same distance by dispersion can be shown to be $t_D = (L^2/D_h^*)$. The ratio of these two times may be used as an index of the combined effects of mass flow and dispersion. Let us define:

$$P = (t_D/t_v) = (vL/D_h) \tag{24}$$

This dimensionless parameter is commonly referred to as the Peclet number, and its value ranges between 0 and ∞.

Small values of P imply increasing significance of dispersion and the converse is true for large values of P. In fact, $P = 0$ represents the case of complete mixing (theoretical maximum dispersion), while $P = \infty$ denotes the case of piston displacement (no dispersion). In other words, with a large P, a square pulse of herbicide applied at the soil surface would leach through the soil profile with minimal spreading, whereas small P values lead to a considerable broadening of the herbicide pulse as a result of dispersion.

In this section, we have seen that two dimensionless parameters, P and R, may be used to summarize the effects of physical processes (mass flow and dispersion) and sorption on herbicide transport in soils. These parameters combine various soil properties (θ, ρ, α, ϵ), herbicide properties (K_D), and flow properties (v and D_h).

B. Limitations of Classical Theories

Attempts to apply the equations presented in the foregoing section for predicting water and solute movement in laboratory soil columns and field profiles have met with varied degrees of success. To facilitate the discussion, the problems encountered in evaluating these theories for herbicide transport may be conveniently grouped into three categories:

1.   Validity of model assumptions.
2.   Inability to provide the required parameters.
3.   Lack of field data to test models.

Each of these aspects will be briefly discussed in the following paragraphs.

A number of simplifying assumptions were made in deriving the differential equations presented in Section II.A. In developing simulation models for predicting herbicide behavior in soils, and especially under field conditions, these transport equations are coupled with other submodels, each with its own set of assumptions and limitations.[25,41,42] Thus, in a field-scale model, the assumptions and process conceptualization may not be valid in one or more of the submodels. The assumed coupling between processes (i.e., interface between submodels) may also be in error. For both these reasons, predicted and measured herbicide concentration distributions may not agree.

It must be recognized that any failings in the Richard's equation to describe water flow will also be reflected in predictions of herbicide flow. In this regard, the presence of macropores (e.g., cracks, fissures, worm holes, root channels, and similar structural features) and preferential flow of water and solutes along these macropores appears to be the major limitation in applying the classical theories for field-scale predictions. The significance of macropore flow is well recognized for inorganic solutes.[43,44]

Rao et al.[45] reported some of the early data demonstrating the impact of preferential flow regions on herbicide leaching during intermittent flooding in field plots of an aggregated

Oxisol. They observed that the peak concentration for picloram herbicide remained near the soil surface even after several irrigations, while small amounts had leached to considerable depths. This was attributed to retention of the herbicide in intraped regions, and a slow diffusive exchange with water flowing preferentially in the interped regions during infiltration.

More recently, Jury et al.[46] have studied the movement of napropamide herbicide movement during unsaturated water flow in a 0.64-ha plot of a sandy loam soil. They reported that after 2 weeks of daily sprinkler irrigation (each irrigation at 0.5 cm hr^{-1}; 23 cm net water applied), 73% of the surface-applied herbicide was retained above the 10-cm depth. However, about 1% of the chemical had penetrated to a depth of 150 cm and trace concentrations could be detected down to the 180-cm depth. Jury et al.[46] suggested that the observed napropamide leaching patterns " . . . could result from macropore flow or from the presence of a mobile soluble complex of napropamide and dissolved organic matter or colloids in suspension." In order to account for the amounts of herbicide detected at deeper depths (about 1%), either the herbicide sorption coefficient (K_{oc}) or the dissolved organic carbon (DOC) content has to be large. The K_{oc} for napropamide is not large (≈350 cm^3/g OC), and the soil organic carbon content of the Tujunga loamy sand is small (OC = 0.59%). Therefore, the formation of a soluble complex of napropamide-DOC is not a likely explanation for the field observations. Hance et al.[47] concluded that movement of fluometuron herbicide associated with soil particles may sometimes represent a small but significant fraction of the total movement, especially over distances of a *few millimeters*. They used ^{144}Ce^{3+} movement as an indicator of soil particle movement. Whether soil particles (and other colloidal particles) could have moved to a depth of 180 cm in the Jury et al.[46] field studies cannot be evaluated, but also may be an unlikely factor. The more plausible explanation then appears to be that of preferential flow, even though strong structural features (e.g., channels, etc.) may not be evident in this soil.

Several earlier workers have also reported that under field conditions, while a major portion of the soil-applied pesticide is retained near the surface (0 to 15-cm depth), a small portion might readily leach to greater depths. Hamaker[48] and Green[49] have reviewed the experimental data on this topic and concluded that slow rate of desorption coupled with intraped solute diffusion might be responsible for the observed pesticide transport in field soils.

Preferential flow and other soil structural heterogeneities may also be responsible for the frequent observation that dispersion coefficients measured in field studies are much larger (an order of magnitude or more) compared with those measured with packed laboratory columns. Whether macropore flow would have a significant impact on solute flow is determined by, among other factors, the antecedent soil-water conditions, soil hydraulic properties, the flow regime, the spatial distribution (density) and the interconnectedness of the macropore sequences, and the rate of solute transfer by diffusion between the frequently referenced soil water regions — the "mobile" and "immobile" soil-water.[43,44,50] In recent years, a number of models have been proposed that explicitly take into account preferential flow in macropores.[43,51-56] These models have been tested to only a limited extent under field conditions, and none with pesticides.

The assumption of linear, reversible, equilibrium sorption (Equation 15) and the use of a simple first-order kinetics model for degradation (Equation 18) have also been questioned. The problems arising from sorption isotherm nonsingularity and the task of quantifying short- and long-term contributions to herbicide sorption by soils were reviewed by Green.[49]

Models that are more complex than Equation 17 may be constructed for herbicide degradation in soils.[57-61] It must be realized, however, that even for a simple first-order model (Equation 17) it is usually difficult to provide the rate coefficient values k_c and k_s for degradation in the two phases. Quantitative aspects of the protective or catalytic effects of sorption on degradation cannot easily be measured.[60,62]

C. Spatial Variability of Model Parameters

The problems associated with spatial and temporal variability in acquiring the necessary data for model input parameters and pesticide concentration distributions in field studies have been examined by Rao and Wagenet.[63] As noted by Peck,[64] because of spatial variability, "...in most fields a very large number of samples will be needed to obtain a reasonable estimate of the mean properties such as water and solute drainage rate." He further cautions that, "...normal sampling programs can be expected to result in only poor estimates of the mean of those properties which exhibit high spatial variance. A corollary is that the probability is low that extreme values of these properties will be included in the investigation of a few sites in the field." Addressing the same issue as it relates to pesticide studies, Rao and Wagenet[63] noted that about 25 samples may be required to identify the probability density distribution and the appropriate statistical moments (mean, variance, etc.) of the population of pesticide concentrations at a given depth. They state that, "Field sampling for (model) validation purposes becomes a cumbersome task if 25 or more samples must be taken at each (soil) depth and time to adequately account for spatial variability. In pesticide fate and efficacy studies, the problems associated with a large number of samples are particularly evident due to constraints associated with (high) costs of analysis for pesticide residues."

A common technique used in pesticide field studies has been to take a large number of soil samples from several locations and composite them into a smaller number of samples for laboratory analysis. The statistical validity of this technique may be questioned, especially if the specific *locations* of each sample and the spatial structure of the field variability of the property of interest are not known.[65] Therefore, without such information, the economic gains made in reducing the number of analyses by sample compositing may well come at the expense of diminished statistical validity of the measured data. Flatman and Yfantis[66] discussed the use of geostatistical techniques to design optimal soil sampling strategies. Further research into this topic, as it applies to pesticide field studies, is warranted.

In contrast to the data available for spatial variability of soil physical properties, only a limited number of studies have been conducted to date to characterize the spatial variability of pesticide sorption and degradation parameters. The available data have been reviewed by Rao and Wagenet[63] and Rao et al.[67] They coined the terms *intrinsic variability* to denote variability inherent in soil properties, and *extrinsic variability* to denote the variability introduced by soil, crop, pesticide, and water management practices imposed on a field. Rao and Wagenet[63] and Rao et al.[67] noted that the intrinsic variability in K_D and k values within a field (a few hectares in area) was small. Thus, the observed variations in field-scale measurements of pesticide concentrations appear to be primarily due to extrinsic variability and also because of spatial heterogeneities in soil hydraulic properties.

Wagenet and Rao[41] have identified the input data requirements for different types of models. High costs have limited the number of detailed field studies in which data useful for model evaluation are collected. Although a large number of herbicide field trials have been conducted, the primary goals of these studies were to assess persistence and efficacy. Unfortunately, in most cases, however, a number of key site-specific measurements essential for model testing were not made. Thus, most of this vast data base remains unusable for model evaluation. In discussing this problem, Leistra[68] commented that, "The accumulation of results from numerous poorly-defined field experiments seems to be a waste of time and skill. A limited number of high-quality field experiments would be of a better use of resources." He states that the purpose of such field studies should be to evaluate expectations based on computer simulations using laboratory and greenhouse data. We concur with Leistra's assessment. A coordinated effort needs to be undertaken to identify the minimum data set needed for model verification and to recommend the necessary protocols for collecting the required data. Such collaborative efforts have already been completed for characterizing herbicide degradation in soils[39] and for bioassay techniques used for obtaining herbicide dose-response data.[69] We are not aware of any such coordinated field studies aimed at collecting data for verification of models for herbicide behavior in soils.

III. SIMPLIFIED FIELD-SCALE MODELING APPROACHES

As discussed in the previous sections, complex simulation models, based on the fundamental processes governing water and solute transport, may be developed. These models require numerical solutions needing considerable computational time. With ever improving computer hardware technology, this constraint is not likely to be significant in the future. However, inability to provide many of the model input parameters will continue to be a major limitation in applying such models to field-scale problems. Green[49] pointed out that, "The more detailed is the specification of physical and chemical mechanisms controlling pesticide transport, the greater will be the task of determining the parameters." This, in fact, has been the major stumbling block in *application* (not development) of complex predictive models for field scale problems.

There remains a need to develop and test simpler models. The philosophy that guides the development of such models is based on two factors: (1) all of the *major* processes influencing herbicide behavior at a field scale must be included; and (2) the input parameters required in these models for validation and subsequent use should be readily available or easily estimated from available data bases. As defined by Rao et al.,[70] the simple models need to be conceptually complete (all processes included), but at a decreased level of complexity, in order that the model is likely to be accessible to more users. This means that the model predictions will be approximate, but could be useful for management-oriented decisions at a field scale.

Several models that are based on the above stated philosophy are presently available. In most of them, water and solute flow submodels are simplified so as to decrease the computational time and, perhaps more importantly, to eliminate the need for soil hydraulic data (soil hydraulic conductivity, K[h]; soil-water characteristic curve, θ[h]). Two distinctly different approaches to solute transport are usually taken. These are: (1) steady-state flow, and (2) piston displacement.

If the soil-water flow is assumed to be at steady-state, then Equation 23 can be used to describe herbicide transport. An example of this approach is the PESTAN model proposed by Enfield et al.[71] The average soil-water flux (q) is inferred from estimates of the difference between total water input (W) and the cumulative evapotranspiration losses (CET) over some time period T. Thus,

$$q \simeq [(W - CET)/T] \tag{25}$$

where T may be taken to be one cropping season or longer. The q value may also be set equal to the annual average recharge rate (LT^{-1}) computed on the basis of water balance computations from long-term weather records. Thus, one advantage of this model is that daily records for rainfall, irrigation, and evapotranspiration are not needed.

The average pore-water velocity (v) required in the solute transport model is estimated as

$$v \simeq (q/\theta^*) \tag{26}$$

where θ^* is an "average" volumetric soil-water content over the time period of interest. Its value is usually assumed to be equal to the "field-capacity" soil-water content (θ_{fc}). However, if $K(\theta)$ data are available, for the given q value a corresponding θ value may be determined and used as the best estimate of θ^* in Equation 26.[71]

The approximate nature of this model is readily apparent; the herbicide pulse is assumed to move through the soil profile in a continuous manner and at an average velocity of v*. The dynamic event-by-event nature of herbicide movement in the root zone is not accounted for in this approach. Thus, this model is not likely to accurately predict herbicide movement within the root zone. Jones et al.[72] have shown that this model underpredicts the leaching of aldicarb nematicide in sandy soils. Below the crop root zone, in the absence of evapotranspiration, soil-water flow may be more closely approximated by steady flow; hence, this model may be useful for prediction of herbicide migration in the vadose zone. Another use of this model is that it can be used to screen or evaluate the relative mobilities of different herbicides.[71,73-75]

The discrete episodic nature of herbicide movement in the root zone, in response to individual rainfall/irrigation events, can be described by several models based on the piston-displacement concept. In a model proposed by Rao et al.[76] the dispersion effects are ignored and the progressive downward leaching of a herbicide pulse is calculated as follows:

$$Z_i = Z_{i-1} + \left[\frac{I_i - I_d}{(\theta_{fc}R)}\right] ; \qquad I_d < I_i \qquad (27)$$

$$Z_i = Z_{i-1} ; \qquad\qquad\qquad I_d \geq I_i \qquad (28)$$

where

$$I_d = \int_0^{Z_i} [\theta_{fc} - \theta(z)]dz \qquad (29)$$

Z_i and Z_{i-1} are the depths (L) at which the herbicide front is located after the i-th and (i-1)-th events; I_i is the amount (L) of rainfall/irrigation *infiltrating* into the soil for the i-th event; I_d is the amount (L) of soil-water deficit resulting from soil-water uptake by plant roots located above the depth Z_{i-1}; θ (z) represents the soil-water content distribution within the root zone; θ_{fc} is the volumetric "field-capacity" soil-water content and is usually taken to be the θ value at h = -0.1 bar; and R is the retardation factor as defined earlier.

For each water input, the herbicide pulse position is calculated using the above equations. Soil-water depletion resulting from evapotranspiration losses between events can be estimated using an appropriate submodel. The model is computationally simple, however, if uniform soil-water extraction throughout the root zone is assumed. This assumption is especially attractive from a practical stand point since in most cases, the exact nature of the root distribution is unknown.

Davidson et al.[77] presented a simple management model for nitrogen based on the above piston displacement concept. Nofziger and Hornsby[78] developed an interactive software package for microcomputers utilizing the above conceptualization of solute transport in soils. This software, named CMIS, facilitates the storage and manipulation of the soil, weather, and herbicide data needed to computer herbicide leaching. It also provides graphic and tabular outputs for easy interpretations of the model predictions. More recently, Nofziger and Hornsby[79] have modified their computer code to consider layered soil profiles.

In order to examine the utility of Equations 27 to 29, the leaching of two herbicides — picloram and atrazine — in a well-drained sandy soil under natural rainfall conditions was simulated with the CMIS model of Nofziger and Hornsby.[78] The soil, herbicide, and environmental data used in the simulations are summarized in Table 1. The progressive downward leaching of the herbicides over the 100-day simulation period is shown in Figure 1. The amounts and the pattern of rainfall are shown in Figure 1A, while the positions of the

Table 1
PARAMETERS USED FOR SIMULATION
OF HERBICIDE LEACHING IN
LAKELAND FINE SAND[a]

Soil properties

Soil series	Lakeland fine sand
Bulk density (ρ)	1.53 g/cm³
"Field-capacity" (θ_{fc})	0.077 cm³/cm³
Organic carbon content (OC)	0.14%
Root depth	100 cm

Herbicide properties

Atrazine	
Sorption coeff. (K_{oc})	163 cm³/g OC
Degradation rate coeff. (k^*)	0.0144 day^{-1}
Retardation factor (R)	5.53
Picloram	
Sorption coeff. (K_{oc})	26 cm³/g OC
Degradation rate coeff. (k^*)	0.005 day^{-1}
Retardation factor (R)	1.72

Water balance

Total rainfall ($W = \Sigma\, I_i$)	66 cm
Total ET (CET)	33 cm

Simulation period

Start	January 2, 1983
End	April 12, 1983
Total time (T)	100 days

[a]   Soil and herbicide properties and rainfall records are as
provided in the program CMIS of Nofziger and Hornsby.[78]

herbicide pulse and a nonadsorbed solute (such as nitrate) are shown for comparison in Figure 1B and 1C.

The impact of several interacting factors on herbicide leaching are illustrated in Figure 1. From Equations 27 to 29, we note that additional herbicide leaching occurs only when a given water input (I_i) exceeds the existing soil-water deficit (I_d), and that only the amount of water in excess of this deficit ($I_i - I_d$) is effective in causing further downward movement of the herbicide pulse. Thus, the quantity $I_e = (I_i - I_d)$ determines whether a specific rainfall/irrigation event may be considered to be a "leaching event". Note that I_e is dependent upon the pattern of rain events, the antecedent soil-water conditions, and the current position of the herbicide pulse. As the herbicide pulse leaches to lower depths, a given rain would be less effective in causing further leaching. For example, several small rains over the 40- to 60-day period were not large enough to be leaching events (Figures 1B and 1C).

The retardation of herbicide leaching due to sorption is explicitly accounted for by R (see Equation 27). The sorption coefficient (K_{oc}) for atrazine is about six times larger than that for the more water-soluble herbicide, picloram (163 vs. 26). The calculated values of R are 1.72 for picloram and 5.53 for atrazine. This difference in R results in picloram being leached out of the rooting depth ($z = 100$ cm) about 40 days after surface application, contrasted to about 96 days for atrazine. Note that a nonadsorbed solute ($R = 1$) would leach beyond the root zone in only about 30 days.

Rao et al.[76,80] and Jones et al.[72] have demonstrated the applicability of this piston displacement approach for predicting fertilizer and nematicide leaching in coarse-textured soils.

The piston flow model described above predicts the *position* of the herbicide pulse, but

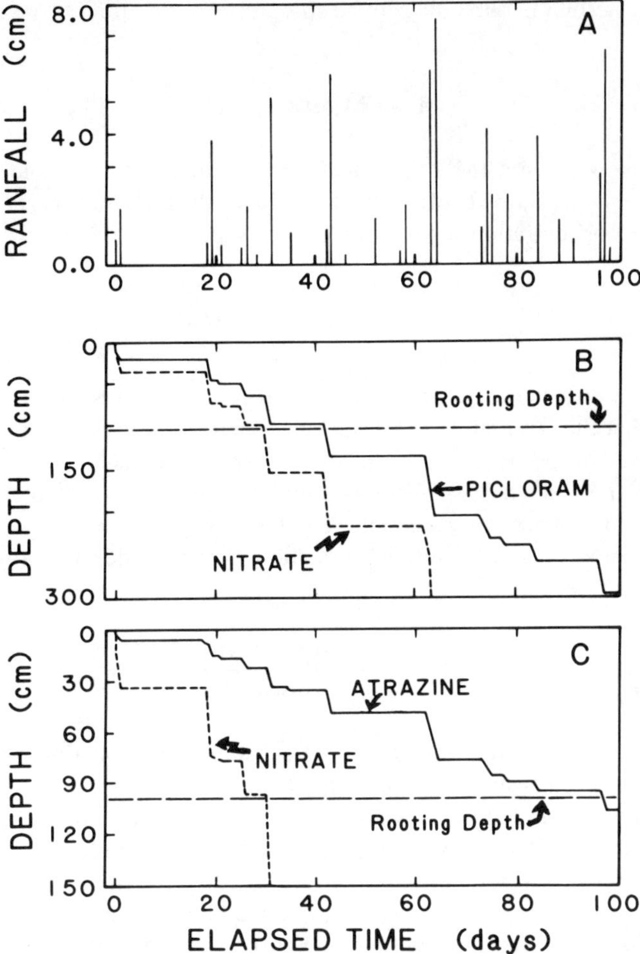

FIGURE 1.   Leaching of two herbicides and nitrate in Lakeland fine sand, simulated by CMIS model (see also text).

not the concentration distribution. The approximate solution, proposed by DeSmedt and Wierenga,[81] Dayananda et al.[82] and Rose et al.[83-86] may be modified to account for sorption and degradation, and used to estimate the herbicide concentration distribution:

$$C(z,t) = \frac{C_o f_t}{2} \left\{ erfc \left[\frac{(z - Z_i)}{2[(D_e/R)t + \epsilon Z_i]^{1/2}} \right] - erfc \left[\frac{(z - Z_i + \delta)}{2[(D_e/R)t + \epsilon Z_i - \epsilon \delta]^{1/2}} \right] \right\} \quad (30)$$

where $C(z,t)$ is herbicide concentration in soil solution (ML^{-3}) as a function of soil depth z and time t; C_o is the herbicide concentration (ML^{-3}) in the soil solution in the initial pulse (t = 0); Z_i is the depth (L) at which the herbicide front is located at time t; δ is the initial width (L) of the herbicide pulse (e.g., depth of incorporation); D_e is the molecular diffusion coefficient ($L^2 T^{-1}$); ϵ is the dispersivity factor (L); R is the retardation factor as defined in Equation 20; f_t is a dimensionless factor that is used to adjust the total herbicide mass to account for losses via degradation; and *erfc* is the complimentary error function. The value of Z_i needed in Equation 30 may be calculated from the piston displacement approach described above (Equations 27 to 29), or other modeling techniques.

In order to estimate the factor f_t in Equation 30, we must first determine the total mass

(M_t; ML^{-2}) of the herbicide remaining in the soil profile at any time t. Assuming first-order decay;

$$M_t = M_o \exp(-k^*t) \tag{31}$$

where M_o is the total amount (ML^{-2}) of herbicide applied at t = 0; and k^* is the depth-averaged first-order degradation coefficient (T^{-1}). Now, the total mass is given by integrating the area under the concentration distribution curve as shown below:

$$M_t = (\theta + \rho K_D) \int_{-\infty}^{+\infty} C(z,t)dz \tag{32}$$

where we have assumed that depth-averaged values for θ, ρ and K_D may be used; and that the $Z_i \gg 0$ such that the integral may be evaluated from $-\infty$ to $+\infty$. (This is similar to the assumptions that soil properties are uniform with soil depth and that the solute peak is located sufficiently below the soil surface. These assumptions are necessary only for computational simplicity. Equation 32 can be numerically integrated otherwise.)

Integrating Equation 32 and substituting Equation 31 for M_t, the following expression for f_t is obtained:

$$f_t \sim \left[\frac{M_o \exp(-k^*t)}{(\theta + \rho K_D)C_o\delta} \right] \tag{33}$$

The f_t value estimated as shown above may be substituted in Equation 30 to calculate the herbicide concentration distribution at any time t.

Note that Equation 30 predicts a Gaussian-like (i.e., bell-shaped) concentration distribution curve for narrow pulse widths (small values of δ). For the example discussed in Figure 1, picloram and atrazine concentration distributions at selected times were calculated using Equation 30. These are shown in Figure 2 as relative herbicide concentrations (C/C_o). We assumed a uniform initial distribution (δ = 10 cm); a dispersivity factor (ϵ) of 1 cm; and D_e = 0.01 cm² day⁻¹. The values of k^* and K_{oc} for the two pesticides are listed in Table 1.

It is evident from the curves shown in Figure 2 that as the herbicide pulse leaches to deeper depths, the maximum concentration (peak height) decreases. This is due to the combined effects of hydrodynamic dispersion, which causes a spreading of the pulse from an initially rectangular shape, and degradation, which results in a decrease in mass (or area under the curve). The less soluble and less persistent herbicide atrazine degrades more rapidly, but moves more slowly through the root zone. At the end of the simulation period (t = 100 days), about 12% of the applied atrazine is found below the root zone. In contrast, about 40% of the picloram is located below the root zone in only 35 days; picloram is more soluble and more persistent compared to atrazine (see Table 1). This demonstrates the combined effects of degradation and sorption on herbicide leaching in soils.

The piston flow model described above is currently being tested with field data collected for aldicarb nematicide leaching in field profiles.[87]

The fundamental premise of the piston flow model is that the infiltrating water completely displaces *all* of the initial water resident in the soil profile. Rao et al.[76] have shown that this assumption may hold for several soils. However, as we have already seen in Section II.B., this assumption may not be valid for structured soils. Addiscott[23,88,89] developed a simple model to account for preferential solute flow in the "mobile" regions, with diffusive solute transfer between the "mobile" and "immobile" soil-water regions. This model was subsequently modified by Nicholls et al.[90] to include soil-water evaporation and herbicide

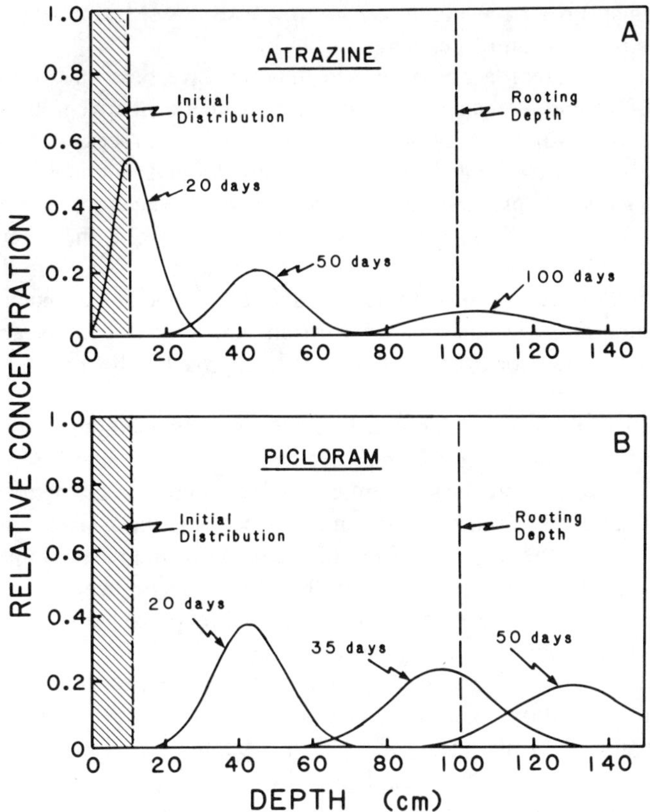

FIGURE 2. Simulated herbicide concentration distributions in Lakeland fine sand.

degradation. They evaluated this modified model with measured data for leaching of two herbicides (atrazine and metribuzin) in fallow field plots. They also evaluated the model proposed by Leistra.[91]

After comparing the measured data with predictions of the two models, and on the basis of difficulties encountered in estimating the model parameters, Nicholls et al.[90] concluded that both models "...simulated soil-water contents with useful accuracy, and were capable of predicting the movement of two herbicides examined sufficiently well for evaluation of their environmental performance in soil." They further note that while the Leistra model was "...a valuable tool for obtaining better theoretical understanding of pesticide behavior...", the Addiscott model "...may be more useful to pesticide technologists, who require a less detailed knowledge of pesticide behavior for specific application." In other words, the simpler model was deemed sufficiently accurate for practical field-scale management application.

Carsel et al.[92] have developed a model for predicting pesticide behavior in the crop root zone. This model, called PRZM (an acronym for pesticide root zone model) is a collection of submodels describing various processes. While a complete review of all these submodels is beyond the scope of this chapter, it should be noted that the soil-water flow submodel offers two options, both based on piston flow concepts. The solute transport model is similar to the convective-dispersive approach discussed in Section II.A. Jones et al.[72] and Carsel et al.[93,94] presented comparisons of field data with PRZM predictions. At the present, data from a number of field studies are being used to evaluate the validity of this model. The

U.S. Environmental Protection Agency currently uses PRZM simulations to help make judgments on the registration of pesticides.[95]

All of the modelling approaches discussed thus far have been *deterministic* rather than *stochastic*. Addiscott and Wagenet[96] note that the distinction between these two modeling techniques is that the former "...presumes that a system or process operates such that the occurrence of a given set of events leads to a uniquely defined outcome," while in the latter approach, the models "...presuppose the outcome to be uncertain and are structured to account for this uncertainty." In other words, for a given field, the deterministic models utilize a single set of parameter values and provide a unique output given these parameters. The stochastic models are designed to accomodate the fact that the model parameters are spatially and temporally variable and as a result, the model outputs are also variable. Wagenet[42] makes the case for the use of stochastic approaches for modelling pesticide fate in soils.

In recognition of soil spatial variability, Green[49] recommended that, "...the best strategy in forecasting pesticide movement for areas one hectare or larger may be to obtain measures of central tendency (mean, geometric mean, etc., depending upon the nature of the statistical distribution) and variances for each key parameter describing the system, and then determine by statistical techniques the likely variation of the depth of the pesticide peak for specified water applications." He further states while this goal may be achieved with an adequate characterization of a field, "...even this level of prediction precision may require more input data than is practical under most pesticide testing programs."

Dayananda[97] has proposed a stochastic version of the simple piston displacement model by considering the variations in rainfall. By assuming that the frequency distribution of rainfall (I_i in Equation 27) fits an exponential distribution, he showed that the probability density function for the solute leaching depth (Z_i) would be a mixture of exponential distributions. Dayananda[97] assumed that all other terms in Equation 27 were constants. It is possible to extend his analysis by considering the water inputs to follow some other frequency distribution (e.g. normal, lognormal, poisson, etc.), and by treating each term in Equation 27 to be a random variable, specified by its own frequency distribution.

Dagan and Bresler[98] have summarized their work on the development of stochastic models, where the spatial variability in soil hydraulic properties is explicitly accounted for. They represent a heterogeneous field by a large number of homogeneous soil "columns", each with its own soil hydraulic characteristics. Within each "column", solute movement by mass flow alone (i.e., piston flow) or convective-dispersive transport was considered. They derive expressions for estimating a field-scale *average* solute concentration, $<C>$, and for the variance (σ^2) about this average value. Their analysis showed that pore-scale diffusion and mechanical dispersion had much less impact on spreading of a solute pulse than did lateral variations in a heterogeneous field.

Jury and co-workers[99-103] have explored the use of transfer function model (TFM) for predicting solute leaching in field soils. As stated by Jury,[99] the TFM, "...has no mechanisms and represents the transport from the surface to a depth L as a stochastic function of the net cumulative amount of water, I, applied at the surface." The TFM approach necessitates the knowledge of the frequency distributions for solute travel times at a specified depth. This is done by calibrating the model with measured field data. Given this, the travel time probabilities at other depths are computed by the TFM model.

In this section we have reviewed the approaches used to simplify the description of water and herbicide transport in soils. These simplified models provide approximate quantitative predictions of persistence and leaching of herbicides in soils. Further testing and refinement of these approaches are warranted, especially for describing pesticide leaching and persistence in structured field soils.

IV. ENVIRONMENTAL SIGNIFICANCE OF HERBICIDE TRANSPORT

Over 1400 active ingredients are found in various pesticide products used to control crop pests in agriculture and silviculture. As noted in other chapters of this book, a number of physical, chemical, and biological processes operating within the crop root zone (usually <2 meters) or in the unsaturated zone (also known as the vadose zone) dissipate these chemicals. It was generally believed that these processes would result in sufficient dissipation of these pesticides, such that the amounts reaching groundwater would be small or at least small enough to be nondetectable and to pose no health risks. This long-held assumption is being questioned as a result of rapid advances in analytical methodology, increasing understanding of chronic toxicological effects of small doses of organic pollutants, and expanded groundwater monitoring efforts. A National Academy of Sciences study[104] concluded that while only about 0.5 to 2% of the U.S. groundwater supplies were contaminated with organic pollutants, much of this contamination seems to occur in areas of heaviest reliance on groundwater.

In a recent paper that summarized the groundwater monitoring data for pesticides in the U.S., Cohen et al.[105] reported that at least 17 pesticides have been detected in groundwater samples collected from a total of 23 states. About half of these chemicals were herbicides (alachlor, atrazine, bromacil, cyanazine, DCPA, dinoseb, metolachlor, metribuzin, and simazine). The reported concentrations for these herbicides ranged from 0.1 to 700 $\mu g/\ell$. Cohen[106] identified 45 pesticides and their metabolites as having a high potential to contaminate groundwater. Of these, 70% were herbicides or growth regulators. This list was expanded to include 90 chemicals that are now proposed for inclusion in a nation-wide groundwater monitoring survey.[107] Nearly half of these chemicals targeted for monitoring are herbicides and their metabolites. Holden[108] has summarized groundwater contamination problems in four states (California, Florida, New York, and Wisconsin) where pesticides have been the focus of considerable public concern and regulatory/monitoring activity. He also provides a listing of pesticides that have been detected in groundwater and those that are suspected to have the potential to contaminate groundwater in these states.

From the foregoing discussion, it is clearly evident that several herbicides, and the by-products of their degradation in soils, have now been identified as potential contaminants of groundwater. Cohen et al.[109] compiled the chemodynamic properties of pesticides detected in groundwater, and concluded that most of these chemicals had aqueous solubility in excess of 30 mg/ℓ, and degradation half-lives longer than 30 days.

The areal extent of the aquifers of interest is generally large enough that a basic constraint to any monitoring program is that samples can only be collected from a limited number of locations. Similar constraints become evident in selecting pesticides to be included in a groundwater monitoring program. Although very sensitive analytical methods have been developed, they are often complex, time-consuming, and expensive. Thus, it may be both impractical and prohibitively expensive to sample an aquifer at a large number of locations, and to analyze these samples for several pesticides. This limitation then forces regulatory agencies to prioritize locations and pesticides and choose a subset from this list. What criteria can be used to prioritize the list of candidate chemicals and the sampling locations such that the limited resources available can be maximized? How can we utilize the existing data and simplified simulation models to assist in establishing the necessary criteria? A considerable effort has been directed at providing an answer to these questions.

Cohen et al.[105] and Lehr et al.[110] discuss the techniques for rating different geohydrologic settings in terms of their *vulnerability* to become contaminated, and offer guidelines for choosing the locations to be included in groundwater monitoring surveys. Their ranking system is based on the DRASTIC index,[111] which is computed from numerical ratings and weights assigned to seven key factors that are thought to determine aquifer vulnerability.

These factors are depth to groundwater, net groundwater recharge rate, nature of the aquifer media, soil type, topography, impact of the vadose zone, and the conductivity of the aquifer. The DRASTIC index was used to produce a map of the U.S. showing the relative vulnerabilities of different hydrogeologic areas.[112]

The *contamination potential* of a pesticide can be evaluated by examining its likelihood to pass through the biologically active root zone. Rao et al.[113] have proposed two indices — the retardation factor (RF) and the attenuation factor (AF) — for ranking pesticides in terms of their *relative* contamination potential. The first index, RF, is based on the travel time (i.e., the time taken to travel past the root zone), whereas the AF index is based on the mass emission (i.e., amount of pesticide leaching past the root zone).

As pesticides leach through the root zone, their rate of movement, compared with that of water, is decreased by sorption on soil. Thus, the differences in pesticide sorption coefficients (K_{oc}) would explain their relative rates of leaching. As already discussed (Equation 20), RF is an index of this difference in leaching rate. An increase in sorption would lead to a larger RF value, which indicates a decreased rate of pesticide leaching. Thus, ranking pesticides by an *ascending order* of RF values produces a list with chemicals having a high contamination (or leaching) potential appearing near the top of the list. This scheme, based on retardation factors, is similar to the mobility rating proposed by Helling and Turner,[114] Helling,[115] and Jury et al.[73-75]

The RF index only considers how fast a pesticide might leach through the root zone for a given water input, and ignores the quantity likely to intrude groundwater. The AF index accounts for pesticide losses resulting from degradation during movement through the root zone. We note that for a given value of half-life, increasing the travel time (or RF) would also increase the losses due to degradation, thus reducing the mass emission. Similarly, for a given RF value (or travel time), larger half-lives would mean larger mass emissions. The value of AF index is computed using information on RF and half-life, and its numerical value is, in fact, equal to the fraction of the applied pesticide that may leach past the root zone. AF values range from 0 to 1; values of 0 and 1 imply that none or all of the applied pesticide is likely to leach past the root zone. Therefore, a list of pesticides by a *descending order* of AF values allows ranking their relative contamination potentials.

Rao et al.[113] evaluated 41 pesticides using these two indices and compared this ranking with that produced by four other schemes. They selected 12 chemicals as having high potential for contaminating groundwater in Florida. Of these, seven were herbicides, three were nematicides, and the other two were insecticides. Rao et al.[113] also discuss the limitations of these two simple indices. Finally, it should be noted that better estimates of the relative travel times and mass emissions can be computed by any one of the simulation models discussed earlier, and site-specific data. Yet input data required are more intensive for such an effort. The main purpose of these indices is only to serve as a general screening tool, and not to provide *predictions* of actual leaching under specific site and environmental conditions.

V. CONCLUSIONS

Nonvolatile herbicides are transported in soil solution as a result of two physical processes: mass flow (or convection) and hydrodynamic disperion. Convection is primarily responsible for the progressive leaching, and determines the position of the herbicide pulse(s). Dispersion results in a broadening of the pulse(s) about the peak. At the laboratory column scale, at least three processes contribute to dispersion: molecular diffusion within a pore, mixing due to pore-velocity distribution, and diffusive solute transfer among pore domains with different velocities. Under field conditions, an additional dominant factor contributing to dispersion is the large-scale heterogeneity in soil hydraulic and chemical properties. Such heterogeneities may arise from the presence of preferential flow paths (cracks, fissures, biochannels, etc.) as well as spatial variability in soil hydraulic properties.

Complex models, based on fundamental laws governing various processes influencing herbicide transport in soils, may be constructed. Such models have been useful in enhancing our understanding of herbicide behavior in soils and for guiding research efforts. However, validation and use of such models at the field scale, has been limited by the lack of detailed data for input parameters and output variables. Simple models may provide "reasonable" predictions of herbicide behavior under field conditions. Several such models have been developed. Validation of even these models has been hampered by lack of reliable field data and collaborative efforts to collect the necessary data.

A number of herbicides have been detected in groundwater. More are likely to be found as systematic groundwater monitoring effects that include more sampling locations and analyses for more pesticides are completed. Simple indices can be developed to evaluate the contamination potential of different herbicides. Such indices can be helpful in selecting chemicals that require careful management and regulation.

ACKNOWLEDGMENT

Portions of this chapter were completed while the senior author was on a sabbatical leave at the Agronomy and Soil Science Department, University of Hawaii, Manoa (UH). Financial support provided by UH is gratefully acknowledged. Approved for publication as Florida Agricultural Experiment Station Journal Series No. 7774.

REFERENCES

1. **Hance, R. J., Ed.,** *Interaction between Herbicides and Soil,* Academic Press, New York, 1980.
2. **Yaron, B., Gerstl, Z., and Spencer, W. F.,** Behavior of herbicides in soils, *Adv. Soil Sci.,* 3, 121, 1985.
3. **Bear, J.,** *Dynamics of Fluids in Porous Media,* American Elsevier, New York, 1971.
4. **Hillel, D.,** *Fundamentals of Soil Physics,* Academic Press, New York, 1980.
5. **Hillel, D.,** *Applications of Soil Physics,* Academic Press, New York, 1980.
6. **Campbell, G. S.,** *Soil Physics with BASIC: Transport Models for Soil-Plant Systems,* American Elsevier, New York, 1985.
7. **Kirkham, D. and Powers, W. L.,** *Advanced Soil Physics,* Wiley-Interscience, New York, 1972.
8. **Hillel, D.,** *Computer Simulation of Soil-Water Dynamics: A Compendium of Recent Work,* International Developmental Research Center, Ottawa, 1977.
9. **Nofziger, D. L.,** Interactive simulation of one-dimensional water movement in soils: user's guide, Circ. No. 675, Institute of Food and Agricultural Sciences, University of Florida, Gainesville, 1985.
10. **Nielsen, D. R., Biggar, J. W., and Davidson, J. M.,** Experimental consideration of diffusion analysis in unsaturated flow problems, *Soil Sci. Soc. Am. Proc.,* 26, 107, 1962.
11. **LaRue, M. E., Nielsen, D. R. and Hagan, R. M.,** Soil water flux below a ryegrass root zone, *Agron. J.,* 60, 625, 1968.
12. **Warrick, A. W., Biggar, J. W., and Nielsen, D. R.,** Simultaneous solute and water transfer for an unsaturated soil, *Water Resour. Res.,* 7, 1216, 1971.
13. **Warrick, A. W., Mullen, G. J., and Nielsen, D. R.,** Scaling field-measured soil hydraulic properties using similar media concept, *Water Resour. Res.,* 13, 355, 1977.
14. **Davidson, J. M., Rao, P. S. C., and Nkedi-Kizza, P.,** Physical processes influencing water and solute flow in soils, in *Chemical Mobility and Reactivity in Soil Systems,* Nelson, D. W. et al., Eds., SSSA Special Publ. No. 11, American Society Agronomy Madison, Wisconsin, 1983, 35.
15. **Lavy, T. L.,** Diffusion of three chloro-*s*-triazines in soil, *Weed Sci.,* 18, 53, 1970.
16. **Scott, H. D. and Phillips, R. E.,** Diffusion of selected herbicides in soil, *Soil Sci. Soc. Am. Proc.,* 36, 714, 1972.
17. **Gerstl, Z., Nye, P. H., and Yaron, B.,** Diffusion of a biodegradable pesticide. I. In a biologically inactive soil, *Soil Sci. Soc. Am. J.,* 43, 839, 1979.

18. **Gerstl, Z., Nye, P. H., and Yaron, B.,** Diffusion of a biodegradable pesticide. II. As affected by microbial decomposition, *Soil Sci. Soc. Am. J.,* 43, 843, 1979.
19. **Biggar, J. W. and Nielsen, D. R.,** Spatial variability of leaching characteristics of a field soil, *Water Resour. Res.,* 12, 78, 1976.
20. **Passioura, J. B. and Rose, D. A.,** Hydrodynamic dispersion in aggregated media. II. Effects of velocity and aggregate size, *Soil Sci.,* 111, 345, 1971.
21. **Rao, P. S. C., Rolston, D. E., Jessup, R. E., and Davidson, J. M.,** Solute transport in aggregated porous media: Theoretical and experimental evaluation, *Soil Sci. Soc. Am. J.,* 44, 1139, 1980.
22. **Rao, P. S. C., Jessup, R. E., and Addiscott, T. M.,** Experimental and theoretical aspects of solute diffusion in spherical and nonspherical aggregates, *Soil Sci.,* 133, 342, 1982.
23. **Addiscott, T. M.,** Simulating diffusion within soil aggregates: a simple model for cubic and other regularly-shaped aggregates, *J. Soil Sci.,* 33, 37, 1982.
24. **van Genuchten, M. Th.,** A generalized approach for modelling solute transport in structured soil, *Mem. Int. Assoc. Hydrogeol.,* 17, 513, 1985.
25. **Rao, P. S. C. and Jessup, R. E.,** Development and verification of simulation models for describing pesticide dynamics in soils, *Ecol. Model.,* 16, 67, 1982.
26. **Rao, P. S. C. and Jessup, R. E.,** Sorption and movement of pesticides and other toxic organic substances in soils, in *Chemical Mobility and Reactivity in Soil Systems,* Nelson, D. W. et al., Eds., SSSA Special Publ. No. 11, American Society of Agronomy, Madison, Wisconsin, 1983, 183.
27. **Hamaker, J. W., and Thompson, J. M.,** Adsorption, in *Organic Chemicals in the Environment,* Vol. 1, Goring, C. A. I. and Hamaker, J. W., Eds., Marcel Dekker, New York, 1972, 49.
28. **Rao, P. S. C. and Davidson, J. M.,** Estimation of pesticide retention and transformation parameters required in nonpoint source pollution models, in *Environmental Impact of Nonpoint Source Pollution,* Overcash, M. R. and Davidson, J. M., Eds., Ann Arbor Scientific, Ann Arbor, Michigan, 1980, 23.
29. **Karickhoff, S. W.,** Semi-empirical estimation of sorption of hydrophobic pollutants on natural sediments and soils, *Chemosphere,* 10, 833, 1981.
30. **Green, R. E. and Karickhoff, S. W.,** Estimation of pesticide sorption coefficients for soils and sediments, in *Supporting Papers for SWAM,* Vol. 3, Agricultural Research Service, U.S. Department of Agriculture, Washington, D.C., 1986.
31. **Walker, A.,** A simulation model for prediction of herbicide persistence, *J. Environ. Qual.,* 3, 396, 1974.
32. **Walker, A.,** Simulation of herbicide persistence in soils. I. Simazine and prometryne, *Pestic Sci.,* 7, 41, 1976.
33. **Walker, A.,** Simulation of herbicide persistence in soils. II. Simazine and linuron in long-term experiments, *Pestic. Sci.,* 7, 50, 1976.
34. **Walker, A.,** Simulation of herbicide persistence in soils. III. Propyzamide in different soil types, *Pestic. Sci.,* 7, 59, 1976.
35. **Smith, A. E. and Walker, A.,** A quantitative study of asulum persistence in soil, *Pestic. Sci.,* 8, 449, 1977.
36. **Walker, A.,** Simulation of the persistence of eight soil-applied herbicides, *Weed Res.,* 18, 305, 1978.
37. **Walker, A. and Zimdhal, R. L.,** Simulation of the persistence of atrazine, linuron, and metolachlor in soil at different sites in the U.S.A., *Weed Res.,* 21, 255, 1981.
38. **Walker, A. and Brown, P. A.,** Spatial variability in herbicide degradation rates and residues in soil, *Crop Prot.,* 2, 17, 1983.
39. **Walker, A., Hance, R. J., Allen, J. G., Briggs, G. G., Chen, Y. L., Gaynor, J. D., Hogue, E. J., Malquori, A., Moody, K., Moyer, J. R., Pestemer, W., Rahman, A., Smith, A. E., Streibig, J. C., Torstenson, N. T. L., Widyanto, L. S., and Zandvoort, R.,** EWRS Herbicide-soil working group: collaborative experiment on simazine persistence in soil, *Weed Res.,* 23, 373, 1983.
40. **van Genuchten, M. Th. and Alves, W. J.,** Analytical Solutions of the One-Dimensional Convective-Dispersive Solute Transport Equation, *Tech. Bull. No. 1661,* Agricultural Research U.S. Department of Agriculture, Service, Washington, D.C., 1983.
41. **Wagenet, R. J. and Rao, P. S. C.,** Basic concepts of modeling pesticide fate in the crop root zone, *Weed Sci.,* 33 (Suppl. 2), 25, 1985.
42. **Wagenet, R. J.,** Principles of modeling pesticide movement in the unsaturated zone, in *Evaluation of Pesticides in Groundwater,* Gardner, W. Y., Honeycutt, R. C., and Nigg, H. N., Eds., ACS Symp. Ser. No. 315, American Chemical Society, Washington, D.C., 1986, 330.
43. **Beven, K. J. and Germann, P.,** Macropores and water flow in soils, *Water Resour. Res.,* 18, 1311, 1982.
44. **White, R. E.,** The influence of macropores on the transport of dissolved and suspended matter through soils, *Adv. Soil Sci.,* 3, 95, 1985.
45. **Rao, P. S. C., Green, R. E., Balasubramanian, V., and Kanehiro, Y.,** Field study of solute movement in a highly aggregated Oxisol with intermittent flooding. II. Picloram, *J. Environ. Qual.,* 3, 197, 1974.

46. **Jury, W. A., Elabd, H., and Resketo, M.,** Field study of napropamide movement through unsaturated soil, *Water Resour. Res.,* 22, 749, 1986.

47. **Hance, R. J., Embling, S. J., Hill, D., Graham-Bryce, I. J., and Nicholls, P.,** Movement of fluometuron, simazine, $^{36}Cl^-$, and $^{144}Ce^{3+}$ in soil under field conditions: qualitative aspects, *Weed Res.,* 21, 289, 1981.

48. **Hamaker, J. M.,** Interpretation of soil leaching experiments, in *Environmental Dynamics of Pesticides,* Haque, R. and Freed, V. H., Eds., Plenum Press, New York, 1975, 115.

49. **Green, R. E.,** Prediction of Pesticide Behavior in the Environment, U.S.A.-U.S.S.R. Symp., *EPA-600/9-84-026,* Office of Research and Development, U.S. Environmental Protection Agency, Athens, Georgia, 1984, 42.

50. **Bouma, J.,** Soil morphology and preferential flow along macropores, *Agric. Water Manage.,* 3, 235, 1980.

51. **Scotter, D. R.,** Preferential solute movement through larger soil voids. I. Some computations using a simple theory, *Aust. J. Soil Res.,* 16, 257, 1978.

52. **Hoogmoed, W. B., and Bouma, J.,** A simulation model for predicting infiltration into cracked clay soil, *Soil Sci. Soc. Am. J.,* 44, 458, 1980.

53. **Skopp, J., Gardner, W. R., and Tyler, E. J.,** Solute movement in structured soils: Two-region model with small interaction, *Soil Sci. Soc. Am. J.,* 45, 837, 1981.

54. **Germann, P. and Beven, K.,** Water flow in soil macropores. I. An experimental approach, *J. Soil Sci.,* 32, 1, 1981.

55. **Beven, K. J. and Germann, P.,** Water flow in soil macropores. II. A combined flow model, *J. Soil Sci.,* 32, 15, 1981.

56. **Beven, K. J. and Clarke, R. T.,** On the variation of infiltration into homogeneous soil matrix containing a population of macropores, *Water Resour. Res.,* 22, 383, 1986.

57. **Hamaker, J. W.,** Decomposition, in *Organic Chemicals in the Environment,* Vol. 1, Goring, C. A. I. and Hamaker, J. W., Eds., Marcel Dekker, New York, 1972, 253.

58. **Hamaker, J. W.,** The application of mathematical modeling to the soil persistence and accumulation of pesticides, in *Persistence of Insecticides and Herbicides,* Monogr. Ser. No. 17, British Crop Protection Council, U.K., 1976, 181.

59. **Goring, C. A. I., Laskowski, D. A., Hamaker, J. W., and Meikle, R. W.,** Principles of pesticide degradation in soils, in *Environmental Dynamics of Pesticides,* Haque, R. and Freed, V. H., Eds., Plenum Press, New York, 1975, 135.

60. **Hurle, K. and Walker, A.,** Persistence and its prediction, in *Interactions Between Herbicides and Soil,* Hance, R. J., Ed., Academic Press, New York, 1980, 83.

61. **Soulas, G.,** Mathematical model for microbial degradation of pesticides in the soil, *Soil Biol. Biochem.,* 14, 107, 1982.

62. **Ogram, A. V., Jessup, R. E., Ou, L. T.,and Rao, P. S. C.,** Effects of sorption on biological degradation rates of (2,4-dichlorophenoxy) acetic acid in soils, *Appl. Environ. Microbiol.,* 49, 582, 1985.

63. **Rao, P. S. C. and Wagenet, R. J.,** Spatial variability of pesticides in field soils: methods for data analysis and consequences, *Weed Sci.,* 33 (Suppl.2), 18, 1985.

64. **Peck, A. J.,** Field variability of soil physical properties, *Adv. Irrig. Sci.,* 2, 189, 1983.

65. **Donigian, A. S., Jr. and Rao, P. S. C.,** Examples of model validation and testing, in *Vadose Zone Modelling of Pollutants,* Hern, S. C. and Melancon, S., Eds., Lewis, Chelsea, Michigan, 1986, 101.

66. **Flatman, G. T., and Yfantis, A. A.,** Geostatistical strategy for soil sampling: the survey and the census, *Ecol. Monit. Assess.,* 4, 335, 1984.

67. **Rao, P. S.C., Edvardsson, K. S. V., Ou, L. T., Jessup, R. E., Nkedi-Kizza, P., and Hornsby, A. G.,** Spatial variability of pesticide sorption and degradation parameters, in *Evaluation of Pesticides in Groundwater,* Garner, W. Y., Honeycutt, R. C., and Nigg, H. N., Eds., ACS Symp. Ser. No. 315, American Chemical Society, Washington, D.C., 1986, 100.

68. **Leistra, M.,** Transport in solution, in *Interactions Between Herbicides and Soil,* Hance, R. J., Ed., Academic Press, New York, 1980, 31.

69. **Nyffeler, A., Gerber, H. R., Hurle, K., Pestemer, W., and Schmidt, R. R.,** Collaborative studies of dose-response curves obtained with different bioassay methods for soil-applied herbicides, *Weed Res.,* 22, 213, 1982.

70. **Rao, P. S. C., Hornsby, A. G., and Jessup, R. E.,** Simulation of nitrogen in agroecosystems: criteria for model selection and use, *Plant Soil,* 67, 35, 1982.

71. **Enfield, C. G., Carsel, R. F., Cohen, S. Z., Phan, T., and Walters, D. M.,** Approximating pollutant transport to groundwater, *Groundwater,* 6, 711, 1982.

72. **Jones, R. L., Rao, P. S. C., and Hornsby, A. G.,** Fate of aldicarb in Florida citrus soils. II. Model evaluation, in *Characterization and Monitoring of the Vadose (Unsaturated) Zone,* Nielsen, D. M. and Curl, M., Eds., National Water Well Association, Worthington, Ohio, 1983.

73. **Jury, W. A., Spencer, W. F., and Farmer, W. J.,** Behavior assessment model for trace organics in soil. I. Description of model, *J. Environ. Qual.,* 12, 558, 1983.

74. **Jury, W. A., Farmer, W. J., and Spencer, W. F.,** Behavior assessment model for trace organics in soil. II. Chemical classification and parameter sensitivity, *J. Environ. Qual.,* 13, 567, 1984.

75. **Jury, W. A., Spencer, W. F., and Farmer, W. J.,** Behavior assessment model for trace organics in soil. III. Application of screening model, *J. Environ. Qual.,* 13, 573, 1984.

76. **Rao, P. S. C., Davidson, J. M., and Hammond, L. C.,** Estimation of nonreactive solute front locations in soils, in Proc. Hazardous Waste Res. Symp., *EPA-600/09-76-015,* Office of Research and Development, U.S. Environmental Protection Agency, Cincinnati, Ohio, 1976, 235.

77. **Davidson, J. M., Graetz, D. A., Rao, P. S. C., and Selim, H. M.,** Simulation of Nitrogen Movement, Transformation, and Uptake in the Plant Root Zone, *EPA-600/3-78-029,* Office of Research and Development, U.S. Environmental Protection Agency, Athens, Georgia, 1978.

78. **Nofziger, D. L. and Hornsby, A. G.,** Chemical movement in soils: *IBM-PC User's Guide,* Circ. No. 654, Institute of Food and Agricultural Sciences, University of Florida, Gainesville, 1985.

79. **Nofziger, D. L., and Hornsby, A. G.,** A microcomputer-based management tool for chemical movement in soils, *Appl. Agric. Res.,* 1, 50, 1986.

80. **Rao, P. S. C., Davidson, J. M., and Jessup, R. E.,** Simulation of nitrogen behavior in the root zone of cropped land areas receiving organic wastes, in *Simulation of Nitrogen Behavior in Soil-Plant Systems,* Frissel, M. J. and van Veen, J. A., Eds., Center for Agricultural Publications and Documentation (PUDOC), Wageningen, Netherlands, 1981, 81.

81. **De Smedt, F. and Wierenga, P. J.,** Approximate analytical solution for solute flow during infiltration and redistribution, *Soil Sci. Soc. Am. J.,* 42, 407, 1978.

82. **Dyananda, P. W. A., Winteringham, F. P. W., Rose, C. W., and Parlange, J. Y.,** Leaching of a sorbed solute: a model for peak concentration displacement, *Irrig. Sci.,* 1, 169, 1980.

83. **Rose, C. W., Hogarth, W. L., and Dayananda, P. W. A.,** Movement of peak solute concentration position by leaching in a nonsorbing soil, *Aust. J. Soil Res.,* 20, 23, 1982.

84. **Rose, C. W., Chichester, F. W., Williams, J. R., and Ritchie, J. T.,** A contribution to simplified models of field solute transport, *J. Environ. Qual.,* 11, 146, 1982.

85. **Rose, C. W., Chichester, F. W., Williams, J. R., and Ritchie, J. T.,** Application of an approximate analytic method of computing solute profiles with dispersion in soils, *J. Environ. Qual.,* 11, 151, 1982.

86. **Rose, C. W., Chichester, F. W., and Phillips, I.,** Nitrogen-15 labeled nitrate transport in a soil with fissured shale stratum, *J. Environ. Qual.,* 12, 249, 1983.

87. **Rao, P. S. C.,** 1986, unpublished data.

88. **Addiscott, T. M.,** A simple computer model for leaching in structured soils, *J. Soil Sci.,* 28, 554, 1977.

89. **Addiscott, T. M.,** Leaching of nitrate in structured soils, in *Simulation of Nitrogen Behaviour of Soil-Plant Systems,* Center for Agricultural Publications and Documentation (PUDOC), Wageningen, Netherlands, 1981, 245.

90. **Nicholls, P., Walker, A., and Baker, R. J.,** Measurement and simulation of the movement and degradation of atrazine and metribuzin in a fallow soil, *Pestic. Sci.,* 12, 484, 1982.

91. **Leistra, M.,** Computed redistribution of pesticides in the root zone of an arable crop, *Plant Soil,* 49, 569, 1978.

92. **Carsel, R. F., Smith, C. N., Mulkey, L. A., Dean, J. D., and Jowsie, P.,** User's Manual for the Pesticide Root Zone Model (PRZM), Release 1, *EPA-600/3-84-109,* Office of Research and Development, U.S. Environmental Protection Agency, Athens, Georgia, 1984, 229p.

93. **Carsel, R. F., Mulkey, L. A., Lorber, M. N., and Baskin, L. B.,** The pesticide root zone model (PRZM): a procedure for evaluating pesticide leaching threats to groundwater, *Ecol. Model.,* 30, 49, 1985.

94. **Carsel, R. F., Nixon, W. B., and Ballantine, L. G.,** Comparison of pesticide root zone model (PRZM) predictions with observed concentraions for the tobacco pesticide metalaxyl in unsaturated soils, *Environ. Toxicol. Chem.,* 5, 345, 1986.

95. **Dean, J. D., Jowsie, P., and Donigian, A. S., Jr.,** Leaching Evaluation of Agricultural Chemicals (LEACH) Handbook, *EPA-600/3-84-0068,* Office of Research and Development, U.S. Environmental Protection Agency, Athens, Georgia, 1984.

96. **Addiscott, T. M. and Wagenet, R. J.,** Concepts of solute leaching in soils: a review of modelling approaches, *J. Soil Sci.,* 36, 411, 1985.

97. **Dayananda, P. W. A.,** A simple model for the path of solute in root zone of soil, *Ecol. Model.,* 16, 81, 1982.

98. **Dagan, G. and Bresler, E.,** Solute transport at field scale, in *Pollutants in Porous Media: The Unsaturated Zone Between Soil Surface and Groundwater,* Yaron, B., Dagan, G., and Goldschmid, J., Eds., Springer-Verlag, New York, 1984, 17.

99. **Jury, W. A.,** Chemical transport modelling: current approaches and unresolved problems, in *Chemical Mobility and Reactivity in Soil Systems,* Nelson, D. W., et al., Eds., SSSA Special Publ. No. 11, American Society of Agronomy, Madison, Wisconsin, 1983, 49.

100. **Jury, W. A., Stolzy, L. H., and Shouse, P.,** A field test of the transfer function model for predicting solute transport, *Water Resour. Res.,* 18, 369, 1982.

101. **Jury, W. A., Sposito, G., and White, R. E.,** A transfer function model of solute transport through soil. I. Fundamental concepts, *Water Resour. Res.,* 22, 243, 1986.
102. **Sposito, G., White, R. E., Darrah, P., and Jury, W. A.,** A transfer function model of solute transport through soil. III. The convection-dispersion equation, *Water Resour. Res.,* 22, 255, 1986.
103. **White, R. E., Dyson, J. S., Haigh, R. A., Jury, W. A., and Sposito, G.,** A transfer function model of solute transport through soil. II. Illustrative applications, *Water Resour. Res.,* 22, 248, 1986.
104. **National Academy of Sciences,** *Groundwater Contamination,* National Academy Press, Washington, D.C., 1984.
105. **Cohen, S. Z., Eiden, C., and Lorber, M. N.,** Monitoring groundwater for pesticides in the U.S.A., in *Evaluation of Pesticides in Groundwater,* Garner, W. Y., Honeycutt, R. C., and Nigg, H. N., ACS Symp. Ser. No. 315, American Chemical Society, Washington, D.C., 1986.
106. **Cohen, S. Z.,** List of potential groundwater contaminants, Internal Memo., Office of Pesticide Programs, U.S. Environmental Protection Agency, Washington, D.C., 1984.
107. **Cohen, S. Z.,** Revised list of analytes for the national pesticide survey, Internal Memo., Office of Pesticide Programs, U.S. Environmental Protection Agency, Washington, D.C., 1985.
108. **Holden, P. W.,** *Pesticides and Groundwater Quality: Issues and Problems in Four States,* National Academy Press, Washington, D.C., 1986.
109. **Cohen, S. Z., Creeger, S. M., Carsel, R. F., and Enfield, C. G.,** Potential for pesticide contamination of groundwater from agricultural uses, in *Treatment and Disposal of Pesticide Wastes,* Krueger, R. F. and Sieber, J. N., Eds., ACS Symp. Ser. No. 259, American Chemical Society, Washington, D.C., 1984, 297.
110. **Lehr, J. H., Aller, L., and Bennett, T.,** Quantifying wisdom: a double-edged sword, *Groundwater Monitoring Review, Spring 1986,* 4.
111. **Aller, L., Bennett, T., Lehr, J. H., and Petty, R. J.,** DRASTIC: A Standardized System for Evaluating Groundwater Pollution Potential Using Hydrogeologic Settings, Office of Pesticide Programs, U.S. Environmental Protection Agency, Washington, D.C., 1985.
112. **Alexander, W. J., Liddle, S. K., Mason, R. E., and Yaeger, W. B.,** Groundwater Vulnerability Assessment in Support of the First Stage of the National Pesticide Survey, Office of Pesticide Programs, U.S. Environmental Protection Agency, Washington, D.C., 1986.
113. **Rao, P. S. C., Hornsby, A. G., and Jessup, R. E.,** Indices for ranking the potential for pesticide contamination of groundwater, *Soil Crop Sci. Soc. Fla. Proc.,* 44, 1, 1985.
114. **Helling, C. S. and Turner, B. C.,** Pesticide mobility: determination by soil thin-layer chromatography, *Science,* 162, 562, 1968.
115. **Helling, C. S.,** Pesticide mobility in soils. II. Applications of soil thin-layer chromatography, *Soil Sci. Soc. Am. Proc.,* 35, 737, 1971.

Chapter 3

HERBICIDES IN SURFACE WATERS

Ralph A. Leonard

TABLE OF CONTENTS

I. INTRODUCTION

Dispersion of pesticide residues into the environment by surface runoff from agricultural lands has been a major concern for about the last 20 years. Early concerns centered primarily on the persistent chlorinated hydrocarbon insecticides which were transported by both water and air far from their site of application.[1,2] Because of their adverse ecological impacts, most of these compounds have been banned or restricted from extensive use. Many pesticides, including herbicides, however, remain in use as essential tools of modern agriculture, and these span a wide range of behavioral and toxicological properties. An analysis of pesticide impacts are beyond the scope of this chapter, but a few observations are useful for perspective. The sensitivity of the public to the pesticide-in-water problem is somewhat a carry-over of concerns that were generated as a result of problems with the persistent organochlorine insecticides. Aquatic life forms were acutely affected at low concentrations and both chronic and acute effects were associated with biological magnification in food chains. Insecticides in current use are much less persistent and many are significantly less toxic. Some of the insecticides, however, are quite toxic to both mammals and aquatic organisms.[3] Significant transport of these materials in runoff from sites of application could, therefore, constitute a hazard. Because most currently used pesticides are nonpersistent and do not concentrate in the food chain, any problem created is likely to be of an acute nature, highly transitory, and occur near the source of runoff.

Herbicides used in agriculture currently constitute the greatest tonnage of all pesticides. Although some herbicides are acutely toxic to fish,[3] the hazard of herbicides to the aquatic environment is small compared to insecticides.[4] With a few exceptions, herbicides have little toxicity to humans, wildlife, and livestock. Herbicides may destroy or suppress aquatic vegetation, producing both desirable and undesirable effects. Kemp et al.[5] and Forney and Davis[6] report growth inhibition of submerged aquatic vegetation in the presence of 60 to 1040 μg/ℓ atrazine, depending on species. Few effects were observed at concentrations of 10 μg/ℓ. The presence of herbicide residues, such as picloram, may also impact the quality of water for irrigation of sensitive crops.[7]

In general, much more data are available on the presence of pesticide residues in surface water than on their environmental significance. As an example, we have amassed data on transitory concentration pulses in streams and in storm hydrographs from plots and watersheds ranging in time from minutes to a few hours. These data have been necessary to understand transport processes and controlling variables. Toxicity tests are usually based on 24-, 48-, and even 96-hr exposure to static concentrations. Perhaps we oversimplify when we compare pesticide concentrations from runoff plots and small watersheds to these toxicity values. However, in many instances, runoff concentrations do not exceed concentrations determined to be toxic using these tests, and here, the possibility of an acute problem may surely be dismissed. Modern instrumental methods can detect pesticide residues at concentrations as low as a few parts per trillion or less. When pesticide residues are present in water, even at these levels, some concern as to unknown or long-term effects are inevitable. Therefore, practices and agricultural production methods that minimize or reduce off-site pesticide transport have been promoted, and much research and education is, therefore, devoted to this end.

This chapter covers processes and factors that determine quantities and distribution of pesticides in surface waters emanating from treated lands, and modeling technology for assessing and predicting herbicide runoff. Although water quality implications are present, no attempt is made to cover ecological and water quality issues with respect to particular situations or herbicides. Most of the published data cited deals specifically with herbicides. However, references to studies on insecticide runoff and other chemicals are included to illustrate basic processes and principles as appropriate.

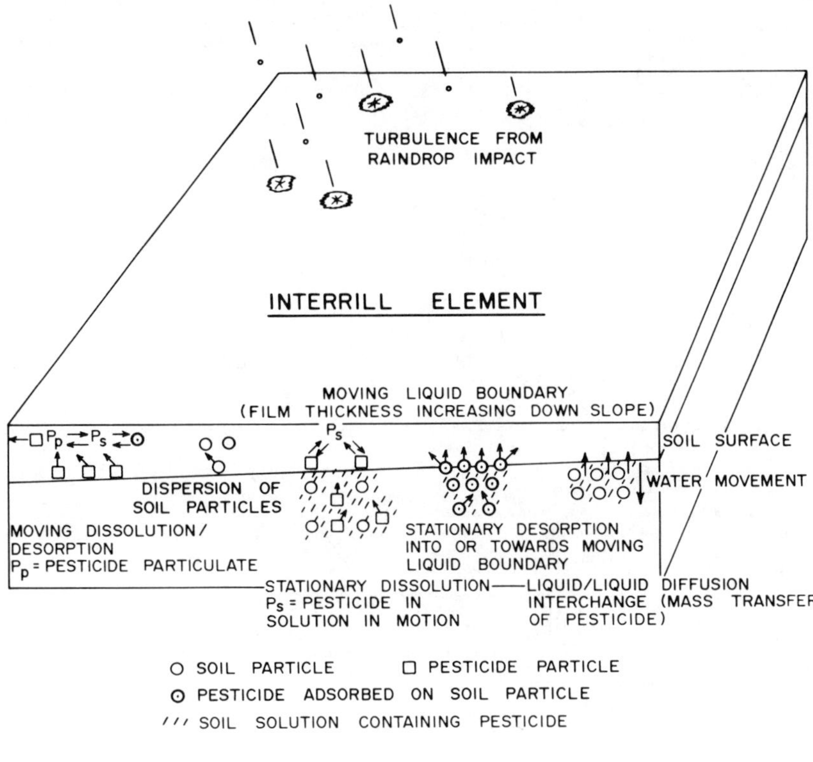

FIGURE 1.   Processes of entrainment and transport of pesticides in surface runoff. (A) Interill element; (B) rill and concentrated flow element.

For the purpose of this chapter, "runoff" is defined as water and any dissolved or suspended matter it contains that leaves a plot, field, or small single-cover watershed in surface drainage. Specifically, "herbicide runoff" includes dissolved, suspended particulate, and sediment-adsorbed herbicide that is transported by water from a treated land surface. At the field-scale, direct surface runoff (overland flow) is the major component of runoff. That is, interflow or return of subsurface water to the soil surface by seepage at this scale is minimal. In some landscapes, larger watersheds will have streams with significant sub-surface or interflow components as well as overland flow components. In these larger, more complex drainage areas, the term "streamflow" will be used instead of runoff.

II. PRINCIPLES GOVERNING ENTRAINMENT AND TRANSPORT IN RUNOFF

A. Extraction into Runoff

Processes of pesticide entrainment and transport in runoff may be visualized, as in Figure 1. At the microscale or interrill scale (Figure 1A) pesticide extraction into runoff may be described as mechanisms of (1) diffusion and turbulent transport of dissolved pesticide from soil pores to the runoff stream; (2) desorption from soil particles into the moving liquid boundary; (3) dissolution of stationary pesticide particulates; and (4) scouring of pesticide particulates and their subsequent dissolution in the moving water.[8] Also, pesticides are entrained in runoff attached to suspended soil particles (sediment) and may be desorbed into the liquid phase as soil particles are momentarily suspended by forces of raindrop impact and scour. Raindrop impact, in addition to suspending and dislodging soil particles, produces

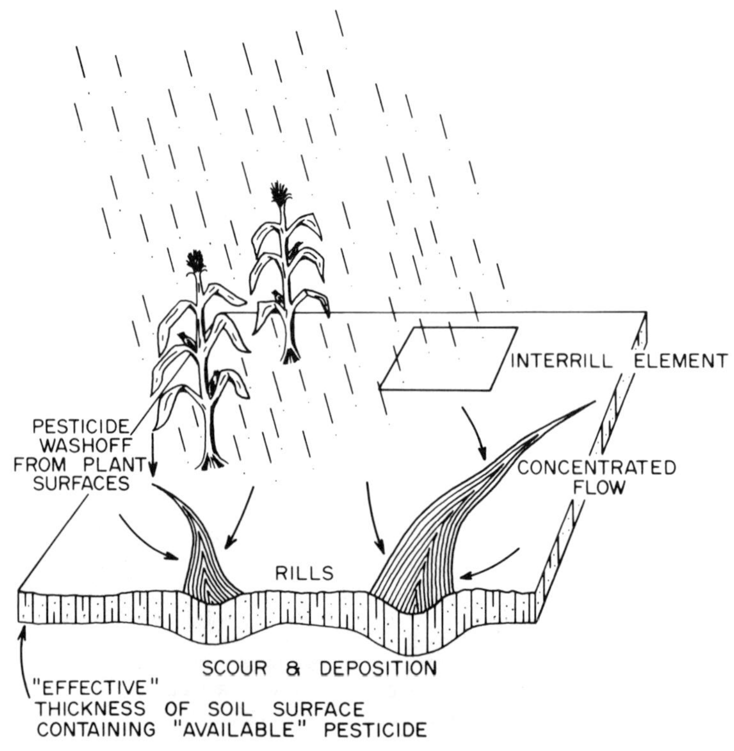

FIGURE 1B.

instantaneous pressure gradients affecting the turbulent interchange between the soil pore solution and the runoff stream.[9,10]

As flows within a field become concentrated because of topography, tillage patterns, or other irregularities affecting flow patterns, rills often develop (Figure 1B). Rills are formed by concentrated flows detaching and transporting soil from a limited part of the land surface.[11] Rills vary in depth and size, but by definition, are limited to depths that can be crossed by farm machinery and filled in by tillage. In contrast to interrill chemical extraction, rill erosion exposes a much deeper layer of the soil surface. Depending on chemical distribution with depth, this soil material may contribute chemicals to the runoff stream or remove chemicals by adsorption.

Shallow interflow may also contribute dissolved chemicals to the runoff stream.[9,12] Shallow interflow is defined as water that has infiltrated the soil surface, but returns to the surface as seepage downslope or into rills, furrows, and other surface depressions.

The process of extraction of chemicals into runoff is obviously complex, with the degree of complexity increasing as the complexity of slope, drainage, and management increases within a catchment. In development of predictive models, modelers have sought a definition of the "effective" surface soil mass or depth that interacts with runoff. Early models assumed the entrainment process to be dominated by soil erosion, and chemical extraction was described as equivalent to a completely mixed reactor, runoff water equilibrating with transported sediment.[13-15] Steenhuis and Walter,[16] Haith and Tubbs,[17] and Williams and Hann[18] used the concept of a "mixing zone" at the soil surface in which chemicals in this zone were assumed to be equal to that in surface runoff. The depth of this mixing zone was conveniently chosen to be 1 cm. Frere et al.[19] did not define a finite thickness of the mixing zone in their model, but did assume complete mixing of soil pore water with runoff water. In their early model, Crawford and Donigian[14] assumed a surface active depth of 0.3 cm,

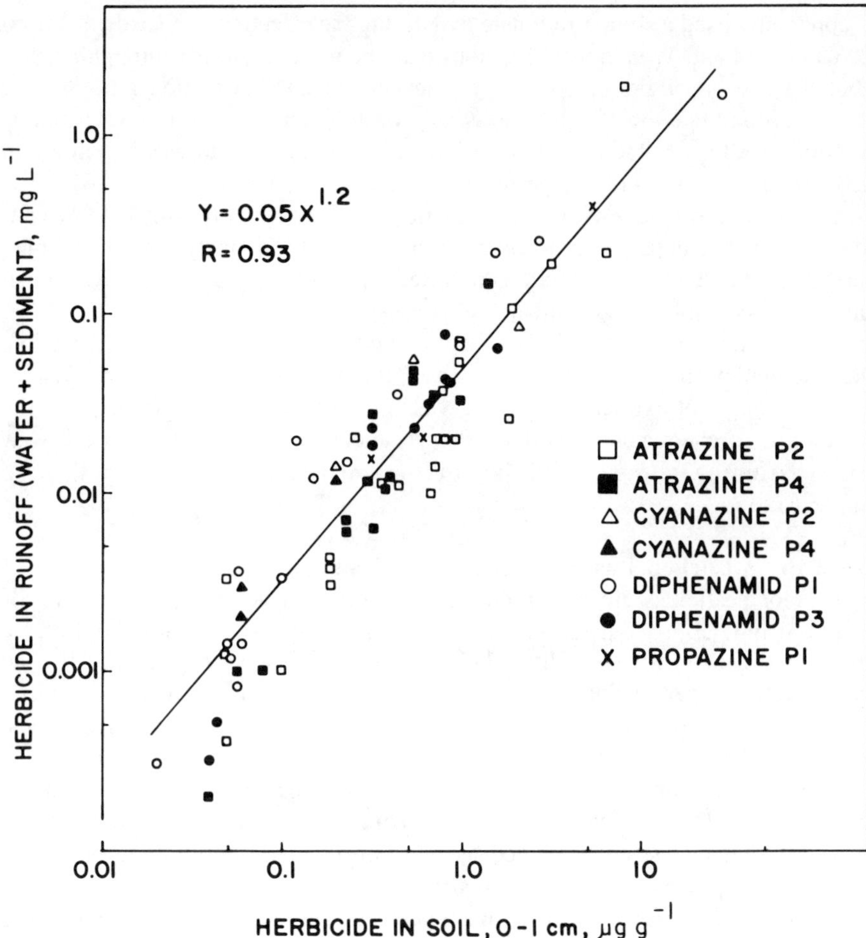

FIGURE 2.   Herbicide concentrations in runoff as related to their concentrations in the 0- to 1-cm surface soil layer.

consistent with later findings of Ahuja et al.[9] Conducting runoff studies using soil boxes and ^{32}P tracers placed at different depths, they concluded that the "effective average" zone of interaction ranged from 0.2 to 0.3 cm. However, the degree of interaction decreased exponentially with depth. Additional studies showed that the effective depth of interaction was related to the degree of soil aggregation and increased with soil slope, kinetic energy of raindrops, and to a lesser extent, rainfall intensity.[20]

Leonard et al.[21] and Smith et al.[22] examined pesticide data from a number of watersheds and found runoff concentrations over a wide range of storm conditions to be strongly correlated with pesticide concentrations in the surface 1 cm of watershed soil (see Figure 2). They, therefore, concluded that this 1-cm surface zone could be used as a descriptor or predictor of the amount of pesticide available for runoff. They made no inferences, however, as to the depth of the layer actually affected by runoff. In development of the CREAMS model, Leonard and Wauchope chose to use the 1-cm depth as a predictor of pesticide available for runoff, primarily because it is about the minimum depth that can be delineated and sampled under actual field conditions considering normal surface roughness.[23] Also, their model is a "lumped" model, where interrill extraction processes and rill processes are not conceptually or mathematically separated and, therefore, the surface depth chosen is an "effective average" in both vertical and horizontal dimensions. Steenhuis and Walter[16] and

Haith[24] apparently used a similar rationale in defining the effective surface depth. In contrast, however, Leonard and Wauchope[23] did not assume complete mixing throughout the source zone, but defined an additional parameter they called an "extraction ratio", which was assumed to approximate the effective mass ratio of the interacting soil to water in the runoff stream. Numerically, a value of 0.05 to 0.2 for the extraction ratio was required to provide reasonable model fit to a wide range of observed watershed and plot data, with a value of 0.1 giving adequate predictions in most situations.[25] Conceptually, this parameter was envisioned as a variable depending on soil properties, rainfall intensity, and runoff rate, much as observed for the depth of interactions reported by Sharpley et al.[20] However, insufficient data have been available for statistical evaluations.

Ingram[26] and Ingram and Woolhiser[27] derived and evaluated a chemical extraction model in which incomplete mixing of pore water and runoff water was assumed. Dispersion of sediment as a source of extracted chemical was not considered. Their work showed that the concentration of a chemical in overland flow is much lower than in soil pore water and, therefore, assuming complete mixing of the surface zone with runoff is a gross oversimplification.

B. Role of the Adsorbed Phase and Sediment Transport

A number of pesticides are preferentially transported, adsorbed to entrained sediments. Wauchope[28] using published data, related mode of transport to pesticide solubility. He found only those pesticides with solubilities below 1 mg/ℓ transported primarily by sediment. Notable exceptions were soluble salts that are strongly adsorbed to clay surfaces, e.g., paraquat, MSMA. Sediment pesticide transport may also be described using adsorption partition coefficients (K_d).[29,30]

Adsorbed pesticide contributes to runoff as sediment-transported pesticide and by pesticide desorption into the flowing water from either moving or stationary soil particles. The mixing zone is a very dynamic system affected by raindrop impact and by the turbulent nature of the overland flow process. In spite of this, most runoff models assume instantaneous and reversible equilibrium.[14,23] The distribution of pesticide between solution and adsorbed phases can then be assumed to follow simple relationships such as the Freundlich equation: $S = KC^N$, where S is the adsorbed-phase concentration, C is the solution-phase concentration, and K and N are constants specific to a pesticide-soil combination. Furthermore, if the concentration ranges of interest are in the nearly linear portion of the adsorption isotherm, $N = 1$, and the relationship reduces to $S = K_dC$. Rao et al.[31] and Green et al.[32] describe in detail equations used for expressing pesticide sorption by soil and methods of estimating sorption constants. Since soil organic matter is the primary soil constituent responsible for adsorption of nonionic organic compounds, the sorption constant may be based on only the organic matter present such that $K_{oc} = K_d/oc$, where oc is the fraction of organic carbon present in the soil or sediment. A sorption constant expressed in this manner is dependent only on the pesticide and is independent of soil type.

For many soil-pesticide systems, adsorption-desorption relationships are very complex and isotherms are often nonsingular and not completely reversible. However, estimated errors in predicting pesticide distribution assuming reversible, singular isotherms are within factors of 2 or 3 for concentration ranges observed in agricultural soil.[33] As indicated above, assuming instantaneous equilibrium may also introduce serious errors in prediction. However, these errors may not be much greater than the uncertainty of 2 or 3 already present. Sorption kinetics are considered in phosphorus entrainment models of Novotny et al.[34] and Sharpley et al.[20] Similar approaches should apply for pesticides and any adsorbed constituent in runoff;[35] however, rate constants may be difficult to evaluate and could depend on soil (sediment): water ratios, pesticide type and concentrations, organic matter contents, and relative ages of the organic matter-pesticide complexes. In spite of the uncertainties, models assuming equilibrium sorption relations appear to give adequate predictions for water quality planning.[36,37] Leonard and Wauchope,[23] however, noted that in some situations the apparent sorption coefficient (K_d) increased with time during a growing season, that is, pesticides appeared to become more difficult to desorb with time of contact in soil. A possible explanation is slow diffusion of pesticide into soil organic matter particles with subsequent desorption becoming kinetically inhibited.

C. Relationships Between Runoff and Distribution and Persistence in the Runoff Zone

Herbicides leaving a treated area in runoff constitute only a small percentage of that applied, since most of the applied herbicide is dissipated by other processes. These processes are discussed at length in other chapters of this publication, and they will be briefly mentioned here only for perspective. An application, depending on crop stage, formulation, intended target, application method, and weather conditions, is distributed between soil, plant foliage or crop residue, and off-catchment losses. Herbicides applied by ground equipment mainly reach the soil surface if that is the intended target. Depending on the density of the weed and crop canopy and application methods, contact herbicides are distributed between the soil surface and crop foliage. Rope-wick applicators and recirculating systems are very efficient and deposit essentially all of the herbicide on foliage.[38] In no-till or conservation tillage systems, herbicides may be intercepted by crop residues.[39] Once a herbicide reaches either the plant or soil surface, it is degraded or transformed by chemical, biological, and photochemical processes and subjected to volatilization losses. Although dissipation is by a combination of processes, pesticide persistence with time on foliage and soil may be approximated by an exponential function mimicking a first-order process.[40,41] Herbicide half-lives or half-concentration times based on an assumed first-order function vary from a few hours to months, depending on the herbicide and the local environment. Herbicide deposited on soil surfaces are subjected to environmental extremes, and therefore, usually dissipate more rapidly than in the soil bulk.[40] Presumably, herbicides residing on foliage and crop residue would also be subjected to many of these same extremes. For rapidly dissipating herbicides, rainfall timing with respect to application is obviously the most critical factor affecting runoff potential.

In a given rainfall/runoff event, the contributions of foliar and plant residue washoff to surface runoff are highly variable, depending on herbicide chemistry and formulation, rainfall timing with respect to application, rainfall intensity and duration, initial soil water content, and other factors that affect water infiltration and runoff. Insecticide washoff and persistence on foliage has been investigated more extensively than herbicide washoff and foliar persistence.[42] Lipophillic compounds, such as organochlorine pesticides, penetrate waxes at the leaf surface and become difficult to dislodge by rainfall. As a result, only 5 to 10% of a compound such as toxaphene is considered available for rainfall washoff, whereas 70 to 80% of the organophosphates may be removed by rainfall.[41] Presumably, the washoff pattern of herbicides residing on foliage and crop residue at the time of a rainfall event is similar. Some triazines[43] and other soluble herbicides are apparently readily removed from crop surfaces.[44,45] Banks and Robinson,[39] however, found that only 45% of applied metribuzin could be washed from wheat straw. Regardless of the fraction of herbicide potentially dislodged, the amount washed off appears to be a function of rainfall volume and independent of rainfall intensity,[46] assuming the same pattern of washoff holds for herbicides as observed with some insecticides. During a rainstorm, concentrations of the organophosphate insecticides methyl parathion and EPN in foliar washoff declined rapidly as hyperbolic functions of rainfall volume. That portion reaching the soil surface before runoff begins may be moved into the soil surface by infiltrating water or become adsorbed by the soil; or in case of dusts and particulates, become intimately mixed with soil. However, pesticides reaching the soil surface before runoff begins may be more susceptible to runoff than soil applied pesticides. Water dripping from plant canopy and crop residue may concentrate pesticides into drip lines, rills, and row furrows which are part of the surface drainage network. After runoff begins, pesticides washed from foliage and plant residue may fall directly into flowing water and be rapidly transported as surface runoff. Runoff begins when rainfall rates exceed infiltration rates. Therefore, even though washoff from the plant canopy is not dependent on rainfall intensity, time to runoff inception may be highly dependent on rainfall intensity and could affect the net contribution of foliar and crop residue washoff to runoff. Conditions most favorable for runoff would be where herbicides are applied to foliage or crop residue immediately before an intense rainstorm when the soil is nearly saturated from a previous rain. Contributions of plant residue washoff to runoff are discussed in more detail in the section on conservation tillage.

Factors that affect runoff potential most are those that affect herbicide persistence in the runoff active zone at the soil surface. These are factors that affect overall herbicide resistance or susceptibility to degradation and volatilization, leaching below the soil surface, and physical redistribution of herbicide residues. Tillage may reduce surface herbicide concentrations by incorporation of surface applications or, in rare situations, increase surface concentrations by exposing herbicide residues from lower soil layers. Most of the factors affecting runoff potential interact in a complex manner. Table 1 lists climatic, soil, pesticide, and management factors that are known to affect herbicide runoff; references to selected research are cited therein.

D. Temporal and Spatial Relationships

If a small unit of the soil surface (similar to that in Figure 1A) is viewed as a simple reactor providing entrained chemicals to runoff in proportion to the mass of chemicals present, pesticide concentrations in runoff should decline during the runoff event as an approximate first-order function. The rate of decline would be affected by the efficiency of the mixing/ extraction process and the rate of movement of chemicals through the mixing zone with infiltrating water. Factors in Table 1, such as pesticide solubility, sorption properties, infiltration rates, rainfall intensity, etc., will affect the rate at which a given chemical leaves the mixing zone. On small runoff plots, particularly when using controlled or simulated rainfall, an "ideal" behavior is observed (Figure 3). That is, runoff concentrations are high initially, and decline exponentially with time in the event. This ideal behavior is more difficult to observe as the complexity of the runoff system increases in both time and space. Studies conducted on field-size watersheds or larger plots under natural rainfall conditions often give pesticide runoff concentrations that vary more than an order of magnitude within a single event, and sometimes in a seemingly random manner.[22] Wauchope[28] concluded that event concentrations for a given chemical are almost unique for each situation, depending on rate of application, storm intensity and timing, field size, etc. Some rational trends are usually observed, however, with runoff concentrations being higher in the early portion of runoff events (Figure 4). Rohde et al.[47] observed an "induction" period resulting in higher storm concentrations of the nematocide ethoprop 5 to 10 min after initiation of runoff. Early runoff could have originated in untreated field borders. Results from small plots are controlled primarily by those processes illustrated in Figure 1A as interrill or microscale processes. As the size and complexity of the drainage area increases, runoff water becomes concentrated in rills, furrows, waterways, and field ditches. If a watershed is conceptually divided into a number of small elements, each giving the "ideal" response, the watershed response at the outlet would be an integration of the individual responses. The shape of the concentration vs. time curve at the watershed outlet would then depend on runoff times of travel from each element, the degree of mixing during transport, and effects of channel processes such as deposition, reentrainment, etc. Considering the complexity of most runoff systems, dif-

Table 1
FACTORS AFFECTING HERBICIDE ENTRAINMENT AND TRANSPORT IN RUNOFF

Factors	Comment	Ref.
Climatic		
Rainfall/runoff timing with respect to pesticide application	Highest concentration of pesticide in runoff occurs in first significant runoff event after application. Pesticide concentration and availability at the soil and foliar surfaces dissipate with time thereafter.	44, 49, 50, 52, 81, 82, 110—112
Rainfall intensity	Surface runoff occurs when rainfall rate exceeds infiltration rate. Increasing intensity increases runoff rate and energy available for pesticide extraction and transport. May also affect depth of surface interaction. Increasing intensity reduces time to runoff within storm.	20, 113
Rainfall/duration/amount	Affects total runoff volumes. Pesticide washoff from foliage related to total rainfall amount. Leaching below soil surface also affected.	42, 44, 67, 110
Time to runoff after inception of rainfall	Runoff concentrations increased as time to runoff decreased. Pesticide concentrations and availability are greater in first part of the event before significant reduction occurs as a result of leaching and incorporation by raindrop impact.	51, 57, 80, 114
Water temperature	Little data available, but increasing temperature normally increases pesticide solubility and decreases physical adsorption.	8, 57
Soil		
Soil texture and organic matter contents	Affects infiltration rates; runoff is usually higher on finer textured soils. Time to runoff is greater on sandy soils, reducing initial runoff concentrations of soluble pesticides. Organic matter content affects pesticide adsorption and mobility. Soil texture also affects soil erodibility, particle transport potential, and chemical enrichment factors.	33, 105, 108, 115
Surface crusting and compaction	Crusting and compaction decreases infiltration rates, reduces time to runoff, and increases initial concentrations of soluble pesticides.	51
Water content	Initial soil water content at beginning of a rainstorm may increase runoff potential, reduce time to runoff, and reduce leaching of soluble chemicals below soil surface before runoff inception.	57, 116, 117
Slope	Increasing slope may increase runoff rate, soil detachment and transport, and increase effective surface depth for chemical extraction.	20, 28, 105
Degree of aggregation and stability	Soil particle aggregation and stability affects infiltration rates, crusting potential, effective depth for chemical entrainment, sediment transport potential, and adsorbed chemical enrichment in sediments.	9, 20, 105
Pesticide		
Solubility	Soluble pesticides may be more readily removed from crop residue and foliage during the initial rainfall or be leached into the soil. However, when time to runoff is short, runoff concentration may be enhanced by increasing solubility.	41, 45, 52, 57, 78, 80
Sorption properties	Pesticides strongly adsorbed in soil will be retained near application site, i.e., possibly at soil surface, and be more susceptible to runoff. Amounts of runoff are then dependent on amount of soil erosion and sediment transport.	55, 75

Table 1 (continued)
**FACTORS AFFECTING HERBICIDE ENTRAINMENT AND TRANSPORT IN
RUNOFF**

Factors	Comment	Ref.
Polarity/ionic nature	Adsorption of nonpolar compounds determined by soil organic matter, ionized compounds, and weak acids/bases affected more by mineral surfaces and soil pH. Lipophilic compounds retained on foliage by leaf surfaces and waxes, whereas polar compounds are more easily removed from foliage by rainfall.	33, 41, 49, 50
Persistence	Pesticides which remain at the soil surface for longer periods of time because of their resistance to volatilization, chemical, photochemical, and biological degradation have higher probability of runoff.	28, 70, 106
Formulation	Wettable powders are particularly susceptible to entrainment and transport. Liquid forms may be more readily transported than granular. Esters less soluble than salts produced higher runoff concentrations under conditions where initial leaching into soil surface is important.	28, 47, 49, 50
Application rate	Runoff concentrations are proportional to amounts of pesticide present in runoff zone. At usual rates of application for pest control, pathways and processes (e.g., sorption and degradation rates) are not affected by initial amounts present; therefore, runoff potential is in proportion to amounts applied.	57, 59, 118
Placement	Pesticide incorporation or any placement below the soil surface reduces concentrations exposed to runoff processes.	28, 47, 49, 50, 59
Management		
Erosion control practices	Reduces transport of adsorbed/insoluble compounds. Also reduces transport of soluble compounds if runoff volumes are also reduced during critical times after pesticide application.	4, 30, 55, 75, 119
Residue management	Crop residues can reduce pesticide runoff by increasing time to runoff, decreasing runoff volumes, and decreasing erosion and sediment transport. However, pesticide runoff may be increased under conditions where pesticides are washed from the crop residue directly into runoff water (high initial soil water, clay soil, intense rainfall immediately after pesticide application).	52, 78, 80—83
Vegetative buffer strips	Buffer strips around treated fields may reduce transport of some pesticides by secondary infiltration, sediment deposition and sorption on plant surfaces and detritis.	72,86
Irrigation	Chemical application by sprinkler irrigation may move soluble pesticides into soil surface and reduce runoff potential. Aerial application of pesticides during periods of flood irrigation greatly increases pesticide runoff in surface drainage.	120, 121

ficulty in interpreting concentration patterns within storm events should be expected. A difficulty is also introduced in sampling storm hydrographs. Infrequent instantaneous samples may give erroneous data. Wauchope et al.[48] discuss in detail experimental procedures necessary to obtain acceptable data on herbicide runoff. Runoff models predicting within storm pesticide runoff concentrations also need to consider how to route and combine flows from each subelement of the catchment. To avoid this complexity, several of the published pesticide runoff models are ''lumped'' models, predicting only event-average concentrations and total pesticide yields in runoff by event or day.[16,23,24]

Event-average pesticide concentrations over a growing season or period of pesticide per-

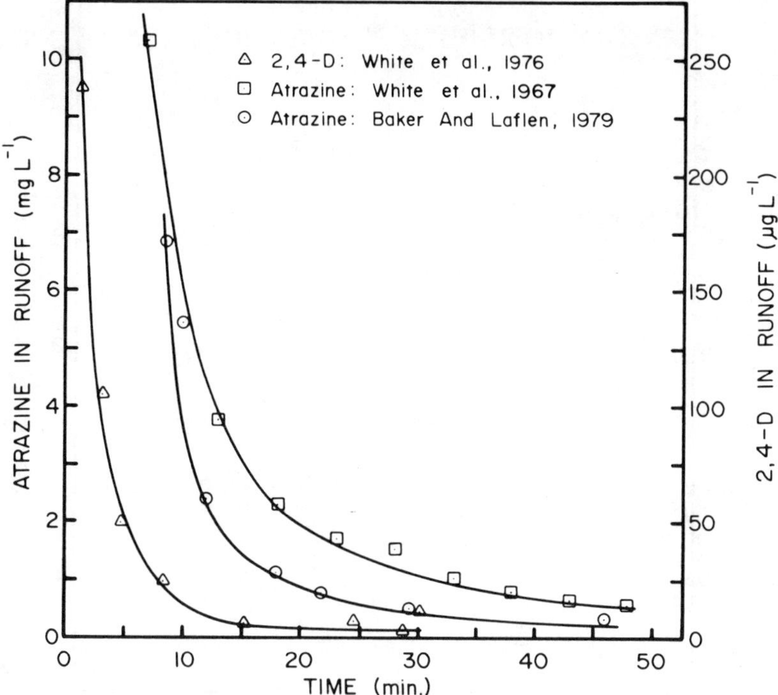

FIGURE 3.   Herbicide runoff concentrations from small plots as related to time after runoff initiation.

sistence are more predictable. Event-average concentrations decrease exponentially with time after pesticide application, as illustrated in Figure 5 (also see Wauchope,[28] Leonard et al.,[21] Wauchope and Leonard[49,50]). This is a reasonable expectation since pesticides are known to dissipate from soil and foliage by processes that can be approximated by first-order rate equations.[40,42] Since runoff concentrations correlate strongly with pesticide concentrations in soil (see Figure 1), we can assume that the decline in runoff concentrations with time reflects primarily a decline at the soil surface.[21] Using semilog plots similar to those in Figure 5, of runoff concentration vs. time, Wauchope and Leonard[49,50] concluded that the "half-lives" of available residues for a wide range of pesticides were remarkably similar, but depended on the time period after application used for calculations. That is, the data only approximated straight-line functions. The curvilinear nature of the plots show that half-lives of the runoff available residues tend to increase over the period of persistence. Stated in another way, the initial pesticide residues, soon after application, that are available for runoff are depleted more rapidly than later in the season. Volatilization, photolysis, and other processes that are driven by the environmental extremes at the soil/air interface presumably contribute to this rapid decline of available pesticide residues. Although dissipation from the soil bulk is approximated by first-order equations, Nash[40] also recognized that dissipation curves could be best described by polynominal functions. Apparently, pesticide half-lives, particularly after one to three half-lives, increase with time in bulk soil somewhat analogous to the observations on runoff available residue, although to varying degrees. With time of contact in soil, pesticides may become somewhat protected from degradation processes by adsorption and/or occlusion.

Physical processes active during runoff may also contribute to the decline in runoff available residues. During the first runoff event after pesticide application, the surface drainage network developed and composed of rills and furrows may be selectively depleted

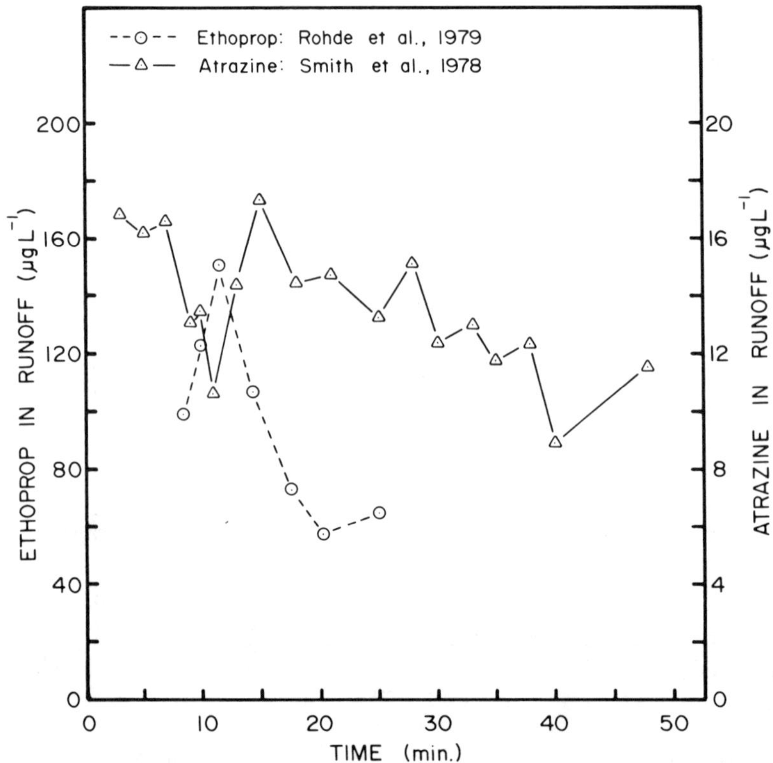

FIGURE 4.   Herbicide concentrations in runoff from farm fields after initiation of runoff.

of pesticide residues. The soil surface may also become armored by sand or gravel as the finer silts and clays are removed. Some pesticide formulations (e.g., wettable powders) may be very easily transported initially due to direct transport of the pesticide carrier. Ease of transport may then decrease as the carrier is mixed into the soil by raindrop impact, overland flow, and tillage.

III. HERBICIDE RUNOFF AND LOADS IN SURFACE WATERS

A. Edge-of-Field Losses

Wauchope[28] conducted an extensive review of available data and proposed a rule-of-thumb pesticide classification with estimates of "average" or "reasonably expected" edge-of-field runoff losses. Estimated seasonal losses were 1% of the amount applied for foliar-applied organochlorine insecticides, 2 to 5% for wettable powders, depending on slope and hydrologic response, and 0.3% or less for the remaining pesticides. Recognizing that in many studies, the bulk of measured runoff occurred in a single storm soon after pesticide application, he defined these as "critical" events; those occurring within two weeks of pesticide application with at least 1 cm rainfall, 50% of which becomes runoff. Noting that runoff losses in the range of 1 to 2% are not uncommon for a wide range of pesticides, he defined "catastrophic" events as those in which runoff losses exceed 2% of the application. These events, almost without exception, are first events after application. Exceptions might be with the most persistent compounds and those, such as paraquat, that are strongly adsorbed and totally dependent on sediment transport.

Herbicide runoff experiments with atrazine, alachlor, trifluralin, picloram, and 2,4-D, summarized in Table 2, support Wauchope's conclusions and illustrate how the many factors

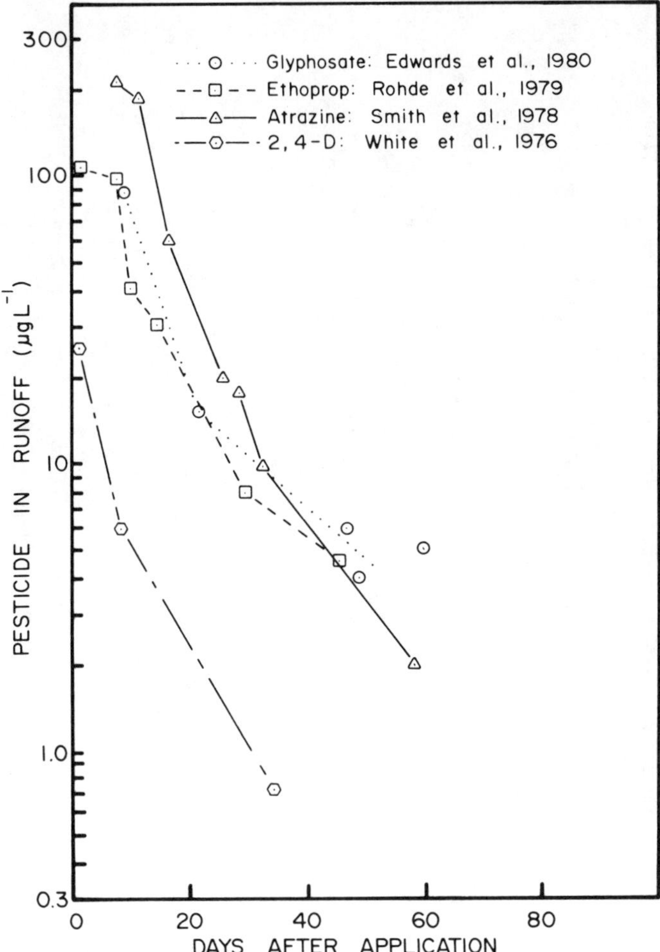

FIGURE 5. Concentration of herbicides in runoff (event averages) in relation to time after application.

interact to determine herbicide transport in surface water. These five compounds represent extremes in herbicide types and formulation, and constitute a major portion of the herbicides applied in agriculture. Data on these five compounds also make up a large percentage of that available on herbicide runoff.

Atrazine is applied as a wettable power to the soil surface. Observed seasonal runoff losses are commonly 2 to 3% of that applied (Table 2). Losses as high as 18% have been reported from small plots in a single intense simulated storm.[51] This magnitude of loss should be approaching a practical upper limit.[49,50] Alachlor is applied as an emulsion, and runoff potential may not be as great as for atrazine. In general, except for the studies of Baker and Johnson,[52] this has been the observation. However, alachlor is not as persistent as atrazine and this lack of persistence alone should decrease seasonal runoff losses. Alachlor is also highly soluble (242 ppm) and perhaps not the best example for emulsifiable concentrates, and this may explain why Baker and Laflen[51] measured as much or slightly more alachlor than atrazine in runoff in their short-term studies. Reported losses of trifluralin are usually 0.5% or less. Trifluralin is a soil-incorporated, emulsifiable concentrate, and resultant concentrations at the soil surface are relatively low. In addition, trifluralin volatilizes readily from the soil surface, rapidly depleting the runoff active zone.[53] Trifluralin is also adsorbed

Table 2
RUNOFF LOSSES OF SELECTED HERBICIDES FROM TREATED PLOTS AND WATERSHEDS

Compound	Location	Type of catchment	Precipitation/irrigation input	Duration and/or variables investigated	General conclusions	Ref.
Atrazine	Watkinsville, Georgia	20-m^2 plots	Simulated rainfall	Single storms rainfall timing and amount	Runoff concentrations highest immediately after application and decreased rapidly within storm	110
	University Park, Pennsylvania	40-m^2 plots 14% slope	Rainfall	Application rates, seasonal relationships	Seasonal losses ranged from 1.7 to 4.7% of application	122
	University Park, Pennsylvania	40-m^2 plots 14% slope	Rainfall	Growing season; mode of transport in runoff	Four to eight times greater loss in water phase compared to sediment	118
	Southern Alabama	20-m^2 plots	Simulated rainfall	Single severe storms; effects of soil type	Soil type affected runoff loss; greatest portion of loss occurred in water phase in first 50 min of runoff	123
	Castana, Iowa	0.5—1.2-ha watersheds	Rainfall	Growing season; total herbicide runoff losses and mode of transport	60% of losses in water phase; seasonal losses from 2.5 to 15.9% reported, but higher value may have been overestimated (see Wauchope[28])	77
	Watkinsville, Georgia	1.3—1.4-ha watersheds	Rainfall	2—3 years duration; conservation practices; data base development for model testing	Seasonal losses ranged from 0.2 to 2% of application and primarily lost in water phase	21
	Coshocton, Ohio	2.7-ha watershed	Rainfall	Growing season; tillage effects	Runoff losses 4.7% of application, dependent on runoff volumes and timing	124
	Coshocton, Ohio	0.4—3.5-ha watersheds	Rainfall	Growing seasons; 3-year study; comparisons of conventional tillage and no till	Average seasonal losses 2% of application; highest loss 6% under most favorable runoff conditions; generally less runoff losses under no-till culture; runoff timing with respect to application most critical	81

Castana, Iowa	Small watershed, 0.55—1.75 ha in size	Rainfall	Growing seasons; 4 years, tillage variables	2.1% average seasonal losses; tillage effects which decreased runoff and erosion decreased herbicides losses, but not in proportion because concentrations in runoff were higher from conservation tillage systems	52
Maryland	Field-size watershed	Rainfall	Growing season; quantify seasonal losses; mode of transport; redistribution in watershed soil	Seasonal losses in runoff about 1%, 60% in water phase	125
Iowa	Small plots	Simulated rainfall	122-mm rainfall applied in 1.75-hr storm; surface application vs. incorporation compared; also effects of compaction by wheel tracks	Total losses were 4.7, 1.6, and 18.3% for surface, incorporated, and surface plus wheel-track compaction, respectively	51
Iowa	Small boxes	Simulated rainfall	70-mm rainfall applied; limed vs. unlimed soil compared	Runoff losses very low from lime soil due to little runoff water; when runoff occurred, losses were 3.7% of application	114
Iowa	14-m^2 plots	Simulated rainfall	127-mm rainfall applied in 2 hr; simulation of 100-year storm; effect of herbicide placement with respect to crop residue investigated	Residue increased time to runoff and therefore reduced herbicide losses; placement had no effect on runoff concentration, but concentrations were negatively correlated with time to runoff	80
Tifton, Georgia	0.02-ha watershed	Rainfall	Two growing seasons; ten plots, two rates	Atrazine appeared in runoff only during first 26 days after application; seasonal losses ranged from 0.22 to 2.24%	126
Southern Ontario	11 agricultural watersheds; 20—70 km^2 in size	Rainfall	Extensive studies on pesticide residues in streamflow, 1975—1977	Overall mean concentration for 11 watersheds was 1.1 and 1.6 µg/ℓ 1976, 1977, respectively; atrazine and simazine were only herbicides detected year-round	66
Lincoln, Nebraska	Dee Creek, a 2025-ha agricultural watershed	Rainfall	3 years	Maximum concentrations in runoff 24 µg/ℓ; limited data base because rainfall only about 60% normal during study period	68

Table 2 (continued)
RUNOFF LOSSES OF SELECTED HERBICIDES FROM TREATED PLOTS AND WATERSHEDS

Compound	Location	Type of catchment	Precipitation/irrigation input	Duration and/or variables investigated	General conclusions	Ref.
	Northwestern Ohio	Honey Creek watershed, a complex agricultural basin	Rainfall	Growing season, 1981	Maximum streamflow concentration of 87 µg/ℓ; 7.5% of applied atrazine in streamflow; rainfall was 2—3 times normal during study period	67
	Wye River, Maryland	Estuary of Chesapeake Bay	Rainfall	Seasonally; 3 years	2—3% of applied atrazine moved to estuary in year with significant runoff within 2 weeks of application; maximum concentration in estuary 15 µg/ℓ at peak loading	63
	Rhode River watershed	Estuary of Chesapeake Bay	Rainfall	3 years; 8 watersheds with varying land use; corn major crop	Atrazine concentration ranged from 0-40 µg/ℓ; no direct relationship between land use and concentratons; atrazine concentrations in runoff higher than that of alachlor, although more alachlor applied to watersheds	64
Alachlor	Castana, Iowa	Small watershed, 0.55—1.75 ha in size	Rainfall 4 years; tillage variables	Growing seasons, 4 years; tillage	About 1% average seasonal runoff losses; conservation tillage reduced runoff volume, but herbicide losses were not reduced in proportion because concentrations in runoff were at times higher under conservation tillage	52
	Maryland, Chesapeake Bay area	Field-sized watershed	Rainfall	Growing season, alachlor losses and redistribution in soil compared to that of atrazine	About 0.2% of application lost in surface runoff compared to about 1% atrazine	125

Iowa, 3 locations	33-m² plots	Simulated rainfall	216-mm rainfall applied in 3 consecutive storms; different amount of residue on plots	Losses of alachlor from all plots were 7.9% and similar for all treatments; reduced herbicide losses due to reduction in runoff and sediment loss was offset by higher herbicide concentrations with residues present	78
Iowa	14-m² plots	Simulated rainfall	122-mm rainfall applied in 1.75-hr storm; surface application vs. incorporation compared; also, effects of compaction by wheel tracks	Total losses were 8.0, 1.7, and 22.1% of application for surface, incorporated and surface plus wheel track compaction, respectively; losses of alachlor were similar to or greater than losses of atrazine	51
Iowa	14-m² plots	Simulated rainfall	127-mm rainfall applied in 2 hr, simulating 100-year storm. Effect of herbicide placement with respect to residue investigated	Residue increased time to runoff, and therefore, reduced herbicide losses; placement had no effect on runoff concentrations, but concentrations were negatively correlated with time to runoff	80
Lincoln, Nebraska	Dee Creek, a 2025-ha agricultural watershed	Rainfall	3 years	Maximum concentration 1.4 μg/ℓ; limited data base because rainfall only 60% of normal during study period	68
Northwestern Ohio	Honey Creek watershed, a complex agricultural basin	Rainfall	Growing season, 1981	Maximum streamflow concentrations of 105 μg/ℓ; rainfall was 2—3 times normal during study period	67
Rhode River watershed	Estuary of Chesapeake Bay	Rainfall	3 years; 8 watersheds with varying land use; corn major crop	Alachlor concentrations ranged from 0—6 μg/ℓ; although alachlor was applied in greater quantity than atrazine, alachlor concentrations in runoff were lower and detectable quantities were found less frequently	64
Southern Ontario	11 agricultural watersheds; 20—70 km² in size	Rainfall	Extensive studies on pesticide residue in streamflow 1975—1977	Ratio of use to loss in streamflow for alachlor estimated to be less than that for atrazine by a factor of 100	66

Table 2 (continued)
RUNOFF LOSSES OF SELECTED HERBICIDES FROM TREATED PLOTS AND WATERSHEDS

Compound	Location	Type of catchment	Precipitation/irrigation input	Duration and/or variables investigated	General conclusions	Ref.
Trifluralin	Eastern North Carolina	17-m^2 plots	Rainfall	6—8 months study to evaluate fate of several chemicals applied in cotton production	Total seasonal runoff losses ranged from 0.3 to 0.5% of application; maximum concentrations observed were 8 to 24 µg/ℓ	127
	Baton Rouge, Louisiana	445-m^2 plots	Rainfall	3 year study to evaluate fate of several pesticides	Total seasonal runoff losses were 0 to 0.05%; concentrations did not exceed 0.5 µg/ℓ	128
	Watkinsville, Georgia	1.3- and 2.7-ha watersheds	Rainfall	Observations seasonally for 2 years on trifluralin; runoff of other herbicides investigated; comprehensive data base for model testing developed	Seasonal losses 0.1 to 0.3% of application; maximum concentration observed, 400 µg/ℓ soon after application	22
	Clarksdale, Mississippi	15.6-ha watershed	Rainfall	Seasonal observations for 2 years reported; other pesticides also included	Seasonal losses were about 0.2%; trifluralin yields in runoff were linearly related to sediment yields in runoff	54
	Tifton, Georgia	0.34-ha watershed	Rainfall plus irrigation (simulated rainfall)	Observations during 2 growing seasons; subplots subjected to intense simulated rainfall	Seasonal losses were 0.03 to 0.17% under natural rainfall conditions; intense simulated rainfall produced losses up to 1.3%; immediate off-site herbicide transport was limited by grassed buffer strips	72
	Clarksdale, Mississippi	15.6-ha watershed	Rainfall	Relationship between sediment yield and pesticide yields developed over 6-year period	Over 6-year period, mean discharge weighted concentration was 0.4 µg/ℓ; trifluralin yield was poorly correlated with annual sediment yield	55
	Imperial Valley, California	28-ha field	Flood irrigation	Studies conducted over 3 years; comprehensive data base developed on pesticides in irrigation runoff	Seasonal losses from cotton fields were 0.29 and 0.14% of application; maximum concentrations in irrigation runoff were 18 µg/ℓ; event mean maximum was 5 µg/ℓ	121

Picloram	Texas	Semiarid rangelands	Rainfall	Picloram persistence and runoff potential	Runoff concentrations were 1 μg/ℓ when runoff occurred more than 30 days after application	56
	Texas	Rangeland	Rainfall	Incidence of picloram in reservoirs	Concentrations in reservoir adjacent to treated areas reached 55 to 180 μg/ℓ when runoff occurred within 2 weeks of application	129
	Texas	9-m^2 plots	Irrigation (simulated rainfall)	Runoff timing and application rates and patterns	Runoff losses 5.5% when entire plot treated; losses 3.2% when only upper half treated	45
	Coshocton, Ohio	0.0008-ha lysimeter	Rainfall	11 months	0.007% of application lost in surface runoff from grass sod	130
	Carlos, Texas	8-ha rangeland drainage area	Rainfall	Spatial and temporal concentration patterns in runoff over a 2-year period	Concentrations in runoff adjacent to treated area declined to 10 μg/ℓ by 12 weeks after application; concentration of 1 μg/ℓ occasionally detected in 1.6 km below area 8 months after treatment	131
	Northwest Texas	3.7—7.2-ha rangeland watersheds	Rainfall and irrigation	Persistence in soils and transport in runoff water for 180 days after application	Concentrations in runoff were 17 μg/ℓ 10 days after application and declined to 1 μg/ℓ 20 to 30 days after application; total losses probably 0.01%	132
	Riesel, Texas	1.2-ha grassed watershed	Rainfall	Runoff concentrations determined during period in which area sprayed 5 times in 6 month intervals	Concentrations were 400 to 800 μg/ℓ if heavy rainfall occurred immediately after treatment, but 5 μg/ℓ if rainfall occurred 1 month or more after treatment; plant washoff was main source of herbicide in runoff	44
	Riesel, Texas	8-ha grassed watershed	Rainfall	Runoff and dissipation-dilution with distance from treated area	Maximum concentrations were 48 and 250 μg/ℓ in initial runoff in 1978 and 1979; concentrations decreased with distance from treated area and with time after treatment; about 6% of application entered runoff during 1-month period when conditions were especially conducive to runoff	71

Table 2 (continued)
RUNOFF LOSSES OF SELECTED HERBICIDES FROM TREATED PLOTS AND WATERSHEDS

Compound	Location	Type of catchment	Precipitation/irrigation input	Duration and/or variables investigated	General conclusions	Ref.
	Cocoino National Forest, Arizona	146-ha Pinyon-Juniper watershed	Rainfall	Persistence and movement of picloram monitored for several years	Maximum runoff concentrations were 320 μg/ℓ in the initial runoff after treatment; picloram not detected farther than 5.6 km downstream; total of 1.1% of the herbicide left treated area	133
2,4-D salt	Watkinsville, Georgia	20-km^2 plots	Simulated rainfall	Runoff from extreme storms; effects of initial soil water content and formulation investigated	After 127-mm rainfall in 2 hr (100-year storm); losses were 3 to 8% from soils initially dry and initially wet, respectively	57
2,4-D	Watkinsville, Georgia	1.3- and 1.4-ha watersheds	Rainfall	Observations for 2 months after application	Seasonal runoff losses were 1.0 and 0.007% of application; maximum concentration 309 and 1.2 μg/ℓ; highest losses from watershed producing runoff soon after application	22
2,4-D salt	Tifton, Georgia	0.34-ha watershed and 30-m^2 subplots	Rainfall plus simulated rainfall	Watershed study conducted for 3 growing seasons under natural rainfall; data supplemented by simulated rainfall on sub plots	Seasonal runoff losses were 0.33, 0.27, and 0.04% under natural rainfall; intense simulated form one day after herbicide application produced 1.5% loss in runoff	134
2,4-D acid	Tifton, Georgia	30-m^2 plots	Simulated rainfall	Study conducted primarily to investigate herbicide transport through adjacent grassed buffer strips	Losses during simulated storm 1 day after application were 10.3% of application from initially wet soil and 2.5% fron initially dry soil	86
2,4-D ester	Watkinsville, Georgia	20-m^2 plots	Simulated rainfall	Runoff fron intense storms; effect of soil water content and formulation investigated	After 127-mm rainfall in 2 hr (100-year) storm losses were 27 and 26% of that applied from soils initially dry and initially wet, respectively	57

| 2,4-D | Southern Oregon | Hill pastures | Rainfall | Brush control and fate of applied herbicides investigated | 2,4-D losses in streamflow were 0.014%; Nearly all of this was contributed from channel banks; no overland flow was apparent | 135 |
| 2,4-D ester | Southwestern Saskatchewan | 4—5-ha plots | Snowmelt runoff | Study conducted over 6-year period to quantify losses of herbicide from treated stubble fields | 6-year average loss was 4.1% of application with a mean flow-weight concentration of 32 μg/ℓ | 136 |

to sediments and in soils with significant adsorption capacity, runoff may be limited by sediment transport.[54,55] Picloram is normally applied as a soluble salt in rangelands and grasslands. Runoff losses reported are highly variable as expected for this type of compound and application. Runoff potential is severely reduced if runoff is delayed 30 days or more after application[44,56] or if rainfall leaches the compound below the soil surface before runoff occurs. Runoff of soluble salts of 2,4-D also reflect this runoff behavior. Barnett et al.[57] measured more than twice as much 2,4-D runoff from initially wet soil than dry soil. Losses of the ester formulation were higher than that of the soluble salt and unaffected by initial soil water content.

B. Herbicides in Streams and Water Bodies

When field runoff containing pesticide enters a water course or body of water, concentrations are rapidly diluted and are also partitioned among various components of the hydrosphere. Herbicides in current use are considered "nonpersistent" compared to the organochlorine insecticides. Early studies, such as those by Nicholson et al.[1,2] concentrated almost totally on defining the problem relative to organochlorine insecticides which were in prevalent use at that time. Results of these and other studies were summarized in reviews by Pionke and Chesters,[58] Leonard et al.,[59] and Caro.[4] In general, widespread contamination by organochlorine compounds, such as DDT and dieldrin, had occurred, although at very low concentrations ranging from a few parts per trillion in water to a few parts per billion in sediments and stream bed material. These concentrations, however low, were deemed environmentally significant because of biological magnification through food chains and resulted in eventual banning of most uses of the organochlorine compounds. Since 1964, the number of reported occurrences and levels of organochlorines in water has continued to decline.[60-62] Very low levels of three herbicides, 2,4-D, 2,4,5-T, and Silvex®, were also occasionally found in rivers and lakes, but at concentrations rarely exceeding a few parts per trillion.

Concerns have been voiced about the possible effects of herbicide residues in surface water on aquatic vegetation. Of particular concern are possible links between atrazine residues and die-back of submerged aquatic vegetation in estuarine systems such as the Chesapeake Bay.[63,64] However, concentrations observed in the Wye River estuary, in a portion of the river receiving high loadings from agricultural runoff, have rarely approached levels thought to produce even minor effects.[63] (see Table 2).

Movement of atrazine in complex watershed systems has been studied extensively in Ontario. Frank and Sirons[65] and Frank et al.[66] reported that between 0.3 and 1.9% of the applied atrazine was transported in streamflow from watersheds ranging in size from 1860 to 7913 ha with peak concentrations of about 33 $\mu g/\ell$. Average concentrations were 1.1 and 1.6 $\mu g/\ell$ in 1976 and 1977, respectively. Atrazine and simazine were two of only three pesticides detected in streamflow throughout the year, but concentrations never reached levels thought to be harmful to vegetation. Baker, Kruger, and Setzler,[67] however, found concentration of atrazine and other herbicides in streamflow from a watershed in northwestern Ohio that could have been sufficiently high enough to produce inhibitory effects on plants and algae (87 $\mu g/\ell$ atrazine, see Table 2). Rainfall during their study period was reported to be two to three times normal, and total atrazine transported was 7.5% of applied. Apparently, runoff conditions during their study met the criteria of "catastrophic" events as defined by Wauchope[28] which must occur rarely on a basin-wide scale.

Atrazine was also reported in streamflow in Nebraska by Schepers et al.[68] and Wu et al.[64] in Maryland, but in much lower concentrations and loads. Residues of 2,4,5-T were found in drainage from the Black Creek watershed in Indiana, but residues of atrazine and alachlor, both of which were used in the watershed, were not detected.[69]

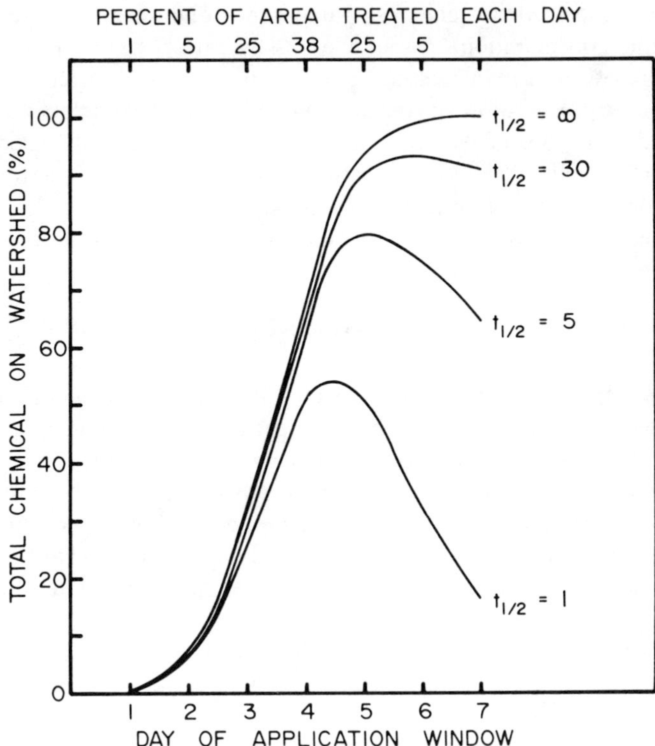

FIGURE 6. Pesticide loads on a hypothetical watershed in relation to temporal application patterns and pesticide persistence.

C. Attentuation in the Transport System

As summarized above, runoff losses at the edge of the field may exceed several percent of the application and concentrations may reach several milligrams per liter if runoff occurs soon after application. As evident from observations in larger diverse watershed systems, these runoff concentrations are rapidly attenuated in the transport system by dilution, deposition and trapping of sediments, adsorption by bottom and bank materials, and infiltration along the various flow paths. Another factor to consider in large diverse watersheds is application timing. Not all the watershed will be treated on any one day. Consider the hypothetical situation depicted in Figure 6. Assume that a given chemical is applied to the entire watershed in a 7-day ''window'', distributed normally about the fourth day. The amounts on the watershed each day potentially available for runoff is represented by a family of curves, depending on the persistence of the chemical applied. The width of the application window could also be varied and it would be obvious that the runoff potential would be reduced as the half-life of the chemical is decreased and/or the width of the window is increased. These concepts may be combined with probability theory, as used by Mills and Leonard,[70] for a quantitative expression of these effects. Also, in large watersheds, rainfall and runoff are usually distributed in time and space such that an additional atentuation of pesticide loads in runoff at the watershed outlet could be expected.

Mayeux et al.[71] demonstrated that a mobile material, such as picloram, may be transported through a watershed system in storm runoff with little loss in actual mass transported, but at concentrations significantly reduced by dilution. In an injection experiment in a stream system, they showed that dilution could reduce concentrations below detectable limits, and net losses in transported mass could be inferred if these limits were not considered. Glotfelty et al.[63] reported a direct proportionality between atrazine concentrations in the Wye River estuary and salinity, and used this relationship to describe dilution and to estimate the average fresh water atrazine concentration (about 23 $\mu g/\ell$) by extrapolation. The work of Trichell et al.[45] illustrates how loss of a mobile herbicide from runoff may occur as a result of infiltration and other processes occurring in overland flow. They treated a plot uniformly with picloram and lost 5.5% of the application in runoff during a simulated rainstorm. By treating the upper one-half of the plot with twice the rate, only 3.2% of the herbicide was lost in runoff. Rohde et al.[72] showed that transport of herbicides, such as trifluralin, may be significantly reduced by infiltration, sedimentation, and filtration processes occurring in grassed waterways and buffer strips.

IV. EFFECTS OF MANAGEMENT ON HERBICIDE RUNOFF

A. Relationships Between Runoff and Erosion Control and Herbicide Runoff

Numerous management practices have been suggested to reduce potential nonpoint source pollution from pesticides.[4,73] These practices include alternative pesticides, crop rotations, integrated pest management systems, conservation practices, substitution of crops, and use of mechanical procedures. Discussions here are limited to interrelationships between soil and water management and pesticide runoff potential. Soil and water management systems may be designed to reduce pesticide runoff where specific problems have been identified, such as runoff from soils containing high levels of persistent residues.[74] However, management systems are usually developed and employed to accomplish other objectives such as reduction of soil erosion. As management is changed, questions are often asked as to the impacts on herbicide runoff potential.

As previously mentioned, the amount of herbicide in the active zone at the soil surface at the time of runoff is perhaps the most important variable affecting amounts and concentrations in runoff. Certainly, changes in rates, formulation/application methods, and chemical type will all affect maximum concentrations and persistence of herbicides available for runoff. Secondary effects are related to changes in infiltration, runoff timing, soil water content, etc. (Table 1). Finally, the effect of management depends on the mode of pesticide transport in runoff and whether or not a given management practice affects runoff volume, sediment yields, or both.

The effects of soil erosion control practices on herbicide runoff depend on the adsorption characteristics of the herbicide and degree of reduction of fine sediment transport. As sediment yield is reduced, adsorbed herbicide in runoff is reduced, but not necessarily in proportion because control practices tend to reduce transport of course particles more than fine particles, and therefore, the capacity for adsorbed chemical transport per unit sediment mass is increased. Sediment enrichment ratios, the ratio of sediment adsorption surface to that of the residual soil, has been reported to range from about 1 to 5 with about 2 to 3 being most common.[22,75,76] However, ignoring these changes in sediment composition and assuming a linear form of the Freundlich equation, the distribution of pesticide between water and sediment can be illustrated as in hypothetical runoff events depicted in Figure 7. Relationships between K_d and the fraction of the total herbicide transported by sediment are shown. The family of curves results from different sediment concentrations. As can be seen, whether a herbicide is transported primarily dissolved in water or attached to sediment depends both on K_d and sediment concentrations. When sediment concentrations are low, solution transport in terms of percent of the total is dominant, even for strongly adsorbed pesticides, because volumes of runoff water greatly outweigh masses of sediment. Conversely, when sediment concentrations are very high, sediment herbicide transport will be significant even for those pesticides with intermediate K_d. As examples, MSMA and paraquat are transported almost entirely by sediment. Whereas, the primary mode of transport of compounds such as trifluralin depends strongly on sediment characteristics and concentrations.[22,28,55]

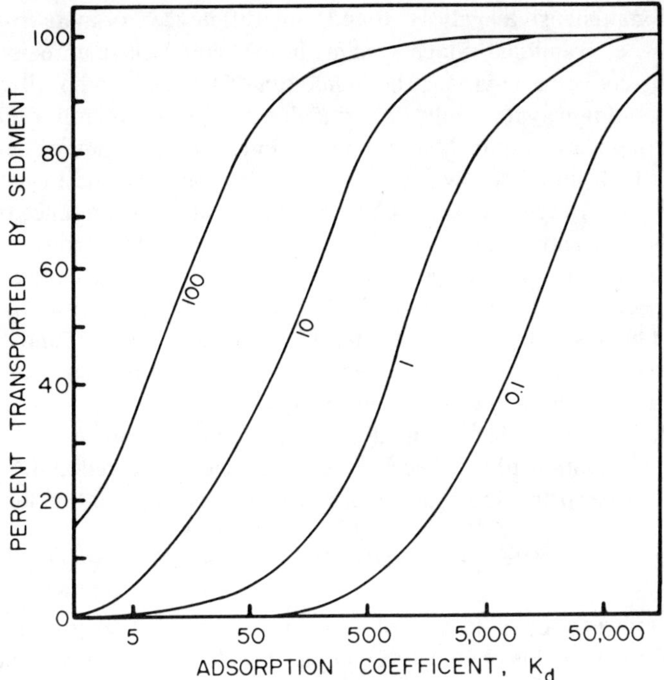

FIGURE 7. Percent of pesticide transported by sediment as functions of pesticide K_d and sediment concentrations ranging from 0.1 to 100 g/ℓ.

Smith et al.[22] compared herbicide runoff from terraced watersheds to runoff from watersheds with no planned conservation practices. Paraquat which was strongly bound to sediment was reduced in runoff in proportion to sediment reduction. Terraces did not reduce runoff volumes, and therefore, losses of atrazine, diphenamid, cyanazine, propazine, and 2,4-D were not affected since these were transported primarily in the aqueous phase. Ritter et al.[77] showed that conservation practices which reduce runoff volumes also reduce losses of atrazine and propachlor. Baker and Johnson[51] and Baker et al.[78] related runoff and soil loss to crop residues in a number of tillage practices. Crop residue reduced runoff volumes on some soils, but not losses of alachlor and cyanazine because concentrations tended to increase with increasing crop residue.

B. Conservation Tillage and Herbicide Runoff

Considerable interest has been expressed in conservation tillage systems and whether pesticide losses in runoff may be enhanced or reduced.[79] Probably both situations exist depending on a combination of conditions. Triazines[43] and other soluble herbicides[44,45] are easily removed from crop surfaces by rainfall and runoff, and this washoff may be a source of enhanced concentrations in runoff as observed by Baker and Johnson[52] and Baker et al.[78] However, Baker et al.[80] reported that concentrations in runoff were not affected by placement of herbicides above or below crop residue, but were negatively correlated with time to runoff. Baker and Laflen[51] had earlier shown that wheel tracks reduced time to runoff, increased initial herbicide concentration in runoff, total runoff volumes, and therefore, total herbicide losses. Triplett et al.[81] and Edwards et al.[82] in their studies of no-till systems emphasized the importance of rainfall/runoff timing relative to herbicide application and reported reduced seasonal herbicide runoff losses as a result of reduced runoff volumes, as did Hall et al.[83] Although herbicides in runoff were not measured, Langdale and Leonard[84] reported significant reductions in growing season runoff under no till systems in the Georgia Piedmont.

Kramer,[85] in a comprehensive analysis, found little difference in runoff from conservation tillage systems vs. conventional tillage systems in the central claypan region of Missouri.

Considering the above observations, herbicide runoff losses from no-till systems may be enhanced under conditions where soluble herbicides are applied to crop residues and runoff occurs soon after application, quickly after rainfall begins. These conditions may occur under a combination of high initial soil water content, intense rainfall, and low infiltration rates (e.g., compacted soil or high clay soil). Conversely, no-till or the presence of crop residues may reduce herbicide runoff under conditions where crop residues reduce surface sealing, maintain higher infiltration rates, and therefore, increase time to runoff and reduce total runoff volumes. Here, most of the soluble herbicide will be leached from the residue to the soil before runoff begins. Langdale and Leonard[84] and Kramer[85] both found runoff volumes, especially from small rainfall events, reduced by conservation tillage during periods of seedbed preparation, when herbicides would be applied.

Runoff of relatively insoluble herbicides and pesticides adsorbed to sediment will be reduced under conservation tillage because sediment yields are reduced.[84] Adsorption or retention on vegetative surfaces may also reduce runoff concentrations of some herbicides.[86]

V. PREDICTION OF HERBICIDES IN SURFACE RUNOFF

Assessing the effects of combinations of weather, management, chemical, and soil factors on pesticide runoff potential is obviously a complex problem. This assessment is made easier by the use of models. Actual measurement of herbicide residues in surface water emanating from treated lands is an expensive and time consuming effort. Experimentally assessing the potential for herbicide runoff from all herbicide-management combinations would be prohibitive. Also, extremely long-term studies would be necessary to adequately sample runoff response under the range of expected climatic events. Modeling and simulation procedures have become extremely valuable in evaluating potential herbicide runoff under a wide range of scenarios.

A. Empirical Methods

Based on data compiled by Wauchope,[28] Wauchope and Leonard[49,50] developed a simple semiempirical formula for predicting maximum concentrations of pesticides in agricultural runoff. Their predicted maximum or ''worst-case'' concentration at any time after pesticide application was

$$C_t = AR(1 + 0.44t)^{-1}$$

where t is time in days after pesticide application, R is the pesticide application rate in kg ha^{-1}, and A is an ''availability index'' with units of ppb ha kg^{-1}. Use of this prediction formula requires classification of pesticides, formulations, and the application situation as to susceptibility of runoff (Table 3). Actually, the ''maximum'' or ''worst-case'' runoff losses may rarely be observed. The absolute maximum concentration for a runoff event is at t = 0 or immediately after a pesticide application. Also, a particular combination of rainfall intensity and duration and soil characteristics is required to produce an event conducive to maximum pesticide runoff. Much of the data used to develop the predictive equation was obtained from experiments using small plots and/or simulated rainfall experiments. Small plots and intense simulated rainfall both may give higher concentrations and loads than observed for large fields. On small plots, runoff may begin sooner and be transported to the sampling point without significant attenuation due to sedimentation and dilution from border areas, etc. On the average, the predictive formula overpredicted by a factor of about 4; however, for its intended use, this was deemed preferable to an underprediction.

Table 3
CLASSIFICATION OF PESTICIDES BY AVAILABILITY INDEX, A[49,50]

Class	A (ppb ha/kg)	Maximum run-off concentration per kg/ha applied	Probable maximum event loss (%)	Pesticide properties and application situation	Example pesticide
I	10,000	10 mg/ℓ	30	Wettable powder applied to soil surface	Cyanazine, prometryne, fluometuron, simazine, atrazine, terbuthylazine, diphenamid, propazine, propachlor, metribuzin, linuron
				Soluble salts applied to soil and strongly bound to clay particles	MSMA, paraquat
				Soluble salts applied to foliage or wet soil	2,4,5-T, 2,4-D, picloram, dicamba
				Pesticides applied in diesel oil	2,4,5-T ester
II	3,000	3 mg/ℓ	10	Soluble salts applied to dry soil	2,4-D, picloram, 2,4,5-T, dicamba, fenac
				Granular and pelleted pesticides regardless of solubility — even if incorporated	Picloram, endrin, fonofos, dieldrin, carbaryl, carbofuran, diazinon
III	1,800	1 mg/ℓ	3	Insoluble pesticides applied to foliage and wet soil	Endosulfan, endrin, DDT, toxaphene, duron, methoxychlor, 2,4-D ester
IV	300	0.3 mg/ℓ	1	Insoluble pesticides applied to dry soil	Alachlor, 2,4-D ester, methoxychlor
				Incorporated wettable powders	Atrazine, dichlobenil
				Incorporated, insoluble, nonpersistent	Trifluralin
				Foliar, insoluble, nonpersistent	Parathion

Adapted from Wauchope, R. D. and Leonard, R. A., *J. Environ. Qual.*, 9(4), 671, 1980.

A probably maximum event loss as a percent of application (Table 3) was developed from the predictive equation as a crude estimate. A runoff event with a maximum concentration of 10 mg/ℓ/kg/ha (a Class I compound) would give 100% loss in 1 cm runoff at this concentration. This concentration could be achieved, however, only in the first portion of the runoff event and would decrease rapidly (see Figure 3). The maximum storm average concentration, therefore, was assumed to be 30% of the maximum. Runoff losses for a single storm may frequently exceed 1 cm, but for these larger storms, the average concentration would be lower.

B. Comprehensive Simulation Models

1. Available Models

In addition to predicting maximum or worst-case pesticide losses for gross assessment of pollution potential, it is often desirable to predict or evaluate how changing management systems or climatic scenarios may affect potential pesticide runoff. More sophisticated modeling techniques have been sought for these purposes which necessitate integrating the conceptual and mathematical description of the pesticide runoff process into hydrologic and soil erosion/sediment transport models.

Crawford and Donigian[14] developed one of the first complete, continuous simulation models they called the Pesticide Transport in Runoff (PTR) model to estimate runoff, erosion, and pesticide losses from field size areas. The hydrologic component of the PTR model is the Stanford watershed model[87] and the erosion component was developed by Negev.[88] The Stanford watershed model was one of the first computer simulation hydrologic models. Donigian and Crawford[89] later incorporated other water quality components into a model called ARM, the agricultural runoff model, which retained the essential features of the PTR model. Continued development of these models and linkages with other water quality models has led to a comprehensive program (HSPF) for modeling runoff, sediment, pesticides, nutrients, and other water quality constituents from urban, agricultural, and other land uses.[90] HSPF allows detailed simulation of stream hydraulics, water quality processes, pesticide and nutrient behavior in soils and water bodies, and sediment contaminant transport. All these modeling programs require extensive data for calibration.

Frere et al.[19] developed an agricultural chemical transport model (ACTMO) to estimate runoff, sediment yield, and chemicals from field- and basin-size areas. The hydrology component is the U.S. Department of Agriculture hydrograph laboratory model[90] which is based on an infiltration concept. The erosion component is based on the rill and interrill erosion concepts and the universal soil loss equation (USLE) modification developed by Foster et al.[92] The ACTMO model requires no calibration with historical data.

Bruce et al.[15] developed an event model to estimate runoff, erosion, and pesticide losses from field size areas for single runoff-producing events. This model requires calibration to the specific site and conditions.

The pesticide runoff model of Haith and Tubbs[17] is structured to couple with any hydrologic and soil loss models. In application, they used the U.S. Department of Agriculture Soil Conservation Service[93] curve number (CN) method for simulation of daily-runoff and the modified USLE as proposed by Williams.[94] They assumed the available pesticide for runoff to be that in the surface 1 cm soil layer, and this pesticide was allowed to decay exponentially with time. A single-valued linear equilibrium adsorption process was assumed for partitioning pesticide between adsorbed and aqueous phases. Since most of the required data inputs are readily available, no calibration is necessary in applying this model.

Steenhuis[95] and Steenhuis and Walter[16] described a model, the Cornell Pesticide Model (CPM), which is similar to the model of Haith and Tubbs[17] in that the Soil Conservation Service CN method and USLE submodels are used as options. Additional hydrology options, however, are included for application to snowmelt and shallow soils. The pesticide submodel has three components: a first-order degradation component, a pesticide leaching component,[96] and a pesticide runoff function. The model uses a mixing depth of about 1 cm, but this depth can be adjusted by calibration using measured pesticide losses. The CPS (continuous pesticide simulation) model tested by Lorber and Mulkey[36] is based on the Cornell model.

The U.S. Department of Agriculture, Agricultural Research Service, in a nationally co-ordinated project, developed and published the CREAMS (Chemicals, Runoff, and Erosion from Agricultural Management Systems) model in 1980.[97] CREAMS is a relatively simple, computer efficient, physically based model for field-size areas. CREAMS does not require observed data for parameter calibration, although observed data are beneficial to adjust the sensitive parameters such that good agreement can be obtained with simulation results. This model is especially useful in making relative comparisons of pollutant loads from alternate management practices. Compared with other pesticide models, unique features are the ability to simulate foliar-applied chemicals and sediment enrichment ratios. Also, complete mixing of the surface layer with runoff water is not assumed.

Application of the above models, with the exception of HSPF, is limited primarily to field-size areas. HSPF was initiated to provide capabilities of continuously simulating dynamics of river basins with generation of detailed information of aquatic impacts and system responses.[90] Other basin and complex-watershed models containing pesticide components are Small Watershed Agricultural Model (SWAM)[98] and the Pesticide Runoff Simulator.[99] Both of these models use basic pesticide concepts from CREAMS with additional algorithms for routing the field outputs through the watershed system. Hydrology and erosion/sediment transport submodels in the Pesticide Runoff Simulator were adapted from Simulator for Water Resources in Rural Basins (SWRRB).[100]

2. Criteria for Model Selection

Models are proven tools for use in assessing water quality impacts and developing management strategies for water quality management.[90,101,102] Model users should carefully consider attributes of different models to determine which can provide the desired results. Some considerations include model purpose, representation, data required, data available, ease of parameter estimation and use, and cost of simulation.

Some models were developed to simulate response for a single, design-type storm. This is entirely adequate when considering only surface runoff and sediment yield, but a design storm may not be one that results in a maximum pesticide loss. For nonpoint source pollutants, continuous (or daily) simulation over a relatively long climatic record is recommended to examine risk analysis.

Often, modelers think that physical processes are the same regardless of scale, and a model can be applied equally as well for fields as for basins. While the processes are the same, the sensitivity of processes vary with size of area. For example, the infiltration process and the simulation time increment in the model are much more significant for small (plot size) areas than for large basins. Therefore, a user should consider the scope of a model when making a selection. Another consideration is whether or not a model is sensitive to changes in management practices. If not, there is little chance of success in selecting among alternate practices for nonpoint source pollution control.

Models that require calibration to evaluate parameter values are generally calibrated for a specific site and practice. If relationships for the physical processes are not carefully formulated, parameter values can be seriously distorted. Calibration of a model with data for a specific site and management practice may give erroneous results when the model is applied to a different site or management practice without recalibration. Therefore, minimization of the need for calibration is desirable. A model is most useful when values for its parameters are readily available as functions of easily measured features of the site and practice being evaluated. Generally, observed data are not available for problem locations, and particularly not for different management practices. If such data are available, then it is not necessary to use a model.

Computer costs may not be important when considering a single model simulation run. However, the model user may be interested in several alternate management practices so a farmer can select the one that controls pollution and is economically feasible within his given constraints. Repetitive 20-year simulations for five to eight alternatives may be prohibitive using inefficient models.

C. Example of Model Applications

A brief description of the CREAMS model and an example application follows. One obvious reason for its selection was because of the author's involvement in its development.[23] Additionally, however, the CREAMS model is well documented,[97] and has received rigorous testing by the developers and other scientists.[36,37]

The hydrology component of the CREAMS model consists of two options depending on availability of rainfall data. Option 1 estimates storm runoff using the CN method as adapted by Williams and LaSeur[103] for simulation of daily runoff when only daily rainfall data are available. Option 2 uses hourly or breakpoint (time-intensity) rainfall data and is based on the Green and Ampt[104] infiltration equation.

The erosion component considers the basic processes of soil detachment, transport, and deposition.[105] The concepts of the model are that sediment load is controlled by the lesser of transport capacity or the amount of sediment available for transport. If sediment load is less than transport capacity, detachment by flow may occur, whereas deposition occurs if sediment load exceeds transport capacity. Raindrop impact is assumed to detach particles regardless of whether or not sediment is being detached or deposited by flow.

The model calculates the sediment transport by sizes and computes a sediment enrichment ratio, based on specific surface area of the sediment and organic matter and the specific surface area for the residual soil. Organic matter, clay, and silt are the principal sediment particles transported which result in high enrichment ratios. The pesticide component estimates concentrations of pesticides in runoff (water and sediment) and total mass carried from the field for each storm during the period of interest. The model accomodates up to ten pesticides simultaneously in a simulation period. Foliar-applied pesticides are considered separately from soil-applied pesticides, because dissipation of pesticides from foliage is often more rapid than from soils. The model considers multiple or repetitive applications of the same chemical as needed. A flow chart of the pesticide component is shown in Figure 8. A 10-mm deep active surface layer is assumed. Movement of pesticides from the surface is a function of runoff, infiltration, and pesticide mobility parameters. Pesticide in the runoff active soil layer is partitioned between the solution phase and the soil phase by the following relationships:

$$(C_w Q) + (C_s M) = aC_p$$

and

$$C_s = K_d C_w$$

where C_w is pesticide concentration in runoff water, Q is volume of water per unit volume of surface active layer, C_s is pesticide concentration in soil that is assumed in equilibrium with runoff, M is mass of soil per unit volume of active surface layer, a is an extraction ratio specifying the soil to water ratio in the extraction zone, C_p is the concentration of pesticide residue in the soil, and K_d is the coefficient for partitioning pesticide between sediment and water phases. Since C_s is the pesticide concentration in the soil material in the runoff zone, selective deposition as expressed by an enrichment ratio enriches this concentration in sediment at the field edge. The amount of pesticide attached to the sediment leaving the field is the product of the concentration C_s, sediment yield, and enrichment ratio.

Pesticide washed from foliage by rain increases the residual pesticide concentration in the soil. The amount calculated as available for washoff is updated between storms by a foliar dissipation function. Pesticide residue in the surface layer is reduced by extraction in overland flow, infiltration, and by degradation described by an exponential function.

Suppose we wish to examine relationships between herbicide persistence, mobility, and

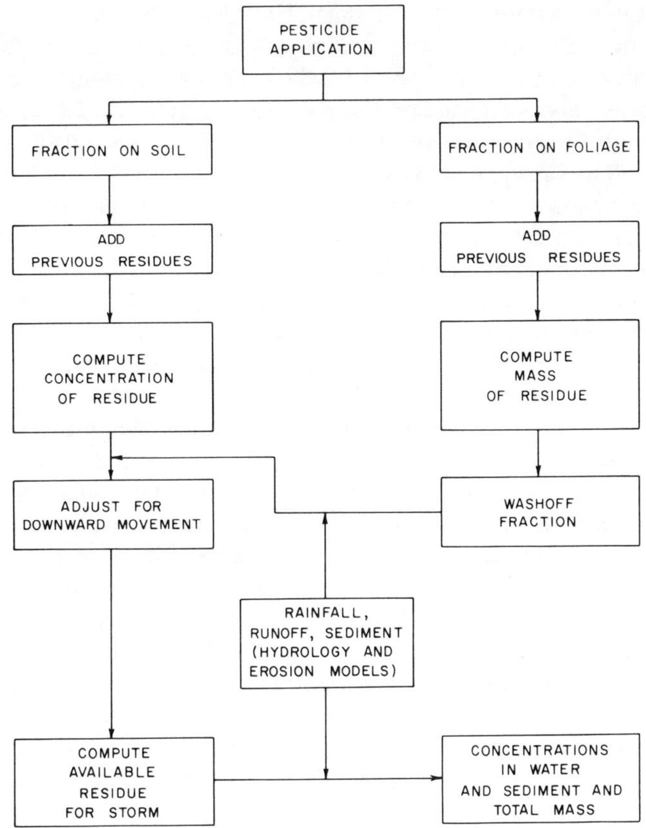

FIGURE 8. Simplified schematic representation of the pesticide sub-model in CREAMS

soil properties, as illustrated by Leonard and Knisel.[106] Four hypothetical herbicides representing a range of properties were envisioned. These were assumed to be applied as postplant sprays after corn planting to a very sandy soil and a clay soil at rates of 3 kg/ha each year for 20 consecutive years. Herbicides A and B with $K_{oc} = 20$ represent nonadsorbed, mobile compounds, whereas C and D with $K_{oc} = 20,000$ represent compounds that are strongly adsorbed by soil and sediments. Herbicides A and C, with half-lives of 3 days, are nonpersistent, whereas B and D, with 30-day half-lives, are moderately persistent. These combinations of mobility and persistence were selected to span a range of classes. A number of pesticides have half-lives longer than 30 days, however, the 30-day value was chosen to avoid any significant carry-over between years of the 20-year simulation period. In practice, corn would not likely be grown for 20 years consecutively, so that each year simulated should be considered as any 1 year of a 20-year period. A 20-year simulation is necessary because shorter simulations do not provide information for frequency analysis. An "average" year or design storm concept has limited value. As will be shown, amounts of herbicides in runoff vary tremendously depending on rainfall patterns, amounts, and timing within any 1 year.

Simulations were conducted using a representative 29-ha field of Tifton loamy sand near Tifton, Georgia, and available rainfall/climatic records for the 20-year period, 1955 through 1974. At Tifton, the long-term average annual rainfall is 1163 mm with a maximum monthly amount of 160 mm in July.[107] Mean annual temperature is 19.2°C with a mean monthly range from 11.1°C in February to 27.2°C in August. Solar radiation ranges from an average

263 ly/day in January to 562 ly/day in June. High-intensity, short-duration, convective thunderstorms commonly occur in the spring and summer during the times of seedbed preparation, planting, and crop emergence, when the soil erosion potential is high. To compare runoff potentials as affected by soil properties, parameters for a clay soil, Houston black clay, were substituted for those for the Tifton soil in a second series of simulations using identical rainfall/climatic records and field size and configuration. Houston black clay occurs in the Blacklands Prairie region in Alabama to Texas, but not in the Tifton area. However, slopes and expected rainfall patterns are not dissimilar. The surface horizon of Tifton loamy sand is composed of medium to very fine sand (over 70% > 50 μm), about 20% silt (2 to 5 μm), less than 10% nonexpanding lattice clay (2 < m), and 1 to 2% organic matter. In contrast, Houston black clay is about 55% clay, 40% silt, an 5% sand with about 2 to 3% organic matter. The Houston soil also contains expanding lattice clays. As a result, the Houston soil has low permeability and hydraulic conductivity compared to the Tifton soil. The Tifton soil is classed as hydrologic group B compared to group D for the Houston soil.[93] Therefore, much higher runoff volumes may be expected from the Houston soil.

Results of simulations for the 20-year period are summarized in Table 4 as annual averages. Recall that CREAMS is not a predictive model, and results are interpreted to show relative differences among soils and pesticides. For the 20-year period, annual rainfall of 1220 mm produced 71 mm runoff from the Tifton soil compared to 386 mm from the Houston soil. Sediment yield from the Houston soil was also greater than that from the Tifton soil in approximate proportion to differences in runoff volume. Because of the greater waterholding capacity of the Houston soil and the higher infiltration rate of the Tifton soil, much greater volumes of water percolated through the root zone of the Tifton soil than through the Houston soil; i.e., 336 mm per year compared to only 0.25 mm per year, respectively. As discussed below, these hydrologic differences had pronounced effects on herbicide runoff.

Herbicide losses expressed as percent of the applications ranged from 0.01 to about 10% (Table 4). These simulated values are consistent with the range of expected values based on actual measurement (see Table 2). Total runoff losses of all four herbicides were less from the Tifton soil than from the Houston soil partly because of lower runoff volumes and sediment losses. However, for the mobile herbicides A and B, (K_{oc} = 20) their movement below the soil surface by water infiltrating through the soil surface before initiation of runoff was the primary factor in reducing runoff of these herbicides from the Tifton soil. This effect is illustrated in Figure 9. Figure 9 shows the concentrations of herbicides at the soil surface available for runoff as computed in the model during a 25-day period in 1973 after herbicide application. In estimating the available surface concentration effective in determining runoff concentrations for a given rainfall event, the herbicide concentration at the soil surface at that time is reduced by the amount transported vertically through the surface 0- to 10-mm layer by infiltrating water. The amount of water infiltrated is assumed to be rainfall less runoff and storage. Since predicted herbicide runoff concentrations are directly proportional to the amount available at the soil surface, this process dramatically reduces runoff potential of mobile herbicides applied to sandy soils with high infiltration rates. As shown in Figure 9, the available surface concentration of herbicides A and B rapidly declined in Tifton soil with rainfall. For example, in 1973, 54 mm of rainfall occurring on the day of application reduced the initial soil surface concentration of herbicides A and B from 20 to 0.14 μg/g. Succeeding rainfall events of 13 and 56 mm on Days 2 and 3 removed essentially all of the remaining herbicides from the soil surface. In contrast to this behavior, the relatively immobile herbicides C and D (K_{oc} = 20,000) remained in the soil surface in both Tifton and Houston soils throughout the 25-day period and declined in concentration as predicted, considering only degradation as described by the herbicide half-life (C = 3 days, D = 30 days). The decline in concentrations of herbicides A and B with time in the Houston soil surface reflected both the effects of infiltration and degradation.

Table 4
SIMULATED RUNOFF, SEDIMENT YIELD, AND LOSSES OF HYPOTHETICAL HERBICIDES[a] FROM A LOAMY SAND AND CLAY SOIL[b]

	Rainfall (mm)	Runoff (mm)	Percolation (mm)	Sediment yield (t/ha)	Herbicide losses		Herbicide sediment phase (% of losses)	Total herbicide losses (%) of application)
					Runoff water (g/ha)	Sediment (g/ha)		
Tifton loamy sand	1220	71	336	10.9				
Herbicide A					0.25	0.001	0.4	0.01
Herbicide B					0.35	0.002	0.5	0.01
Herbicide C					4.07	17.880	81.5	0.73
Herbicide D					12.66	62.650	83.2	2.51
Houston black clay	1220	386	0.25	55.6				
Herbicide A					158.00	0.750	0.5	5.29
Herbicide B					298.00	1.690	0.6	9.99
Herbicide C					5.74	29.040	83.5	1.16
Herbicide D					21.25	125.900	85.6	4.94

[a] Herbicides A and B; K_{oc} = 20. Herbicides C and D; K_{oc} = 20,000.
Herbicides A and C; $t_{1/2}$ = 3 days. Herbicides B and D; $t_{1/2}$ = 30 days.

[b] Annual averages for 20-year period using 1955—1974 rainfall data from Tifton, Georgia.

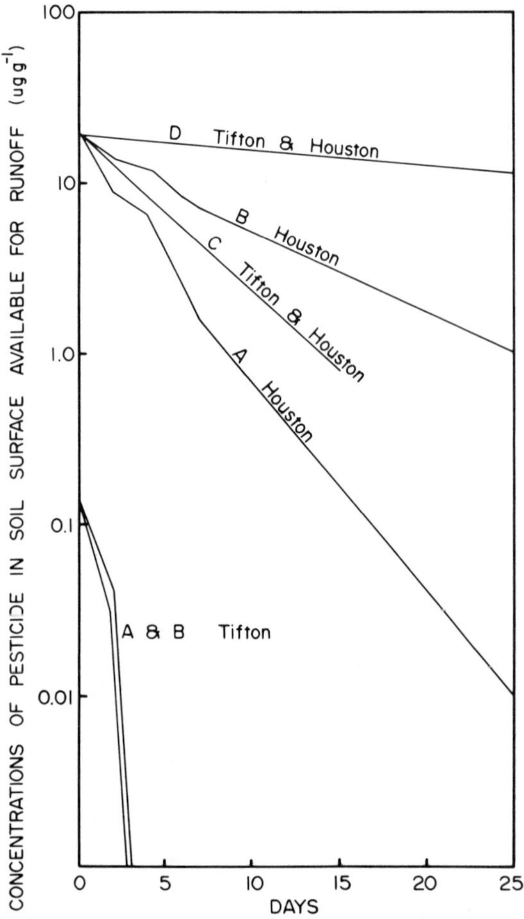

FIGURE 9.   Simulated concentration of herbicides in the surface layer (0 to 10 mm) of Tifton and Houston soils, 1973.

When a herbicide has an inherent high mobility in soil, it may also be readily extracted by surface runoff under conditions where infiltration rate is relatively low compared to runoff rate. These conditions prevailed for herbicides A and B in the Houston soil. Average annual runoff losses were about 5 and 10% for herbicides A and B, respectively (Table 4). Greater losses of B compared to A were incurred because B has a much longer half-life than A.

Herbicides C and D, which are strongly adsorbed by soil organic matter, were transported primarily by sediment from both soils (Table 4). Since 80 to 85% of the total herbicide transport was by sediment, herbicide runoff losses from a given site were proportional to sediment yields. However, annual sediment yields as estimated by simple models, such as the USLE, cannot be used as an index of adsorbed herbicide losses.[108] As an example, annual sediment yields from the Houston soil were about five times that from the Tifton soil, but annual herbicide losses in sediment from the Houston soils were only about twice that from the Tifton soil. This can be explained by considering the timing of sediment losses with respect to herbicide applications. During the spring growing season, immediately after the herbicide application, much of the rainfall occurred in intense thunderstorms; whereas, in fall and winter, rainfall was more of the frontal type and had a lower intensity. Runoff and sediment transport from both soils were more similar during the herbicide application season; whereas, under conditions of frontal rainfall in winter months, much greater runoff

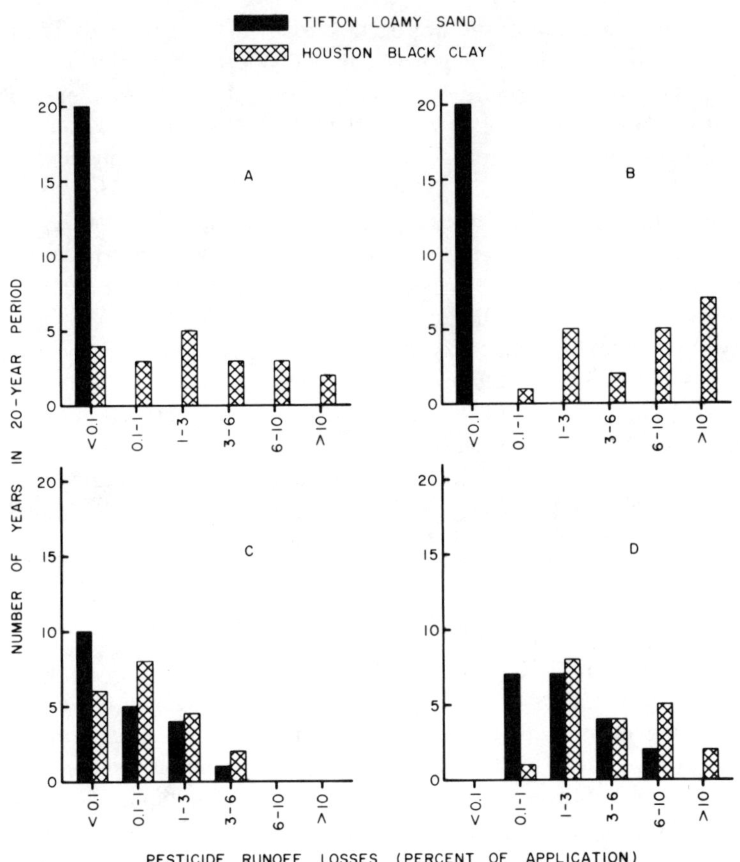

FIGURE 10. Frequency distribution of annual herbicide losses from Tifton and Houston soils.

and sediment transport occurred on the Houston soil. During winter months, most of the applied herbicides had already dissipated and were not present in the transported sediment.

Sediment composition also affects herbicide transport. The CREAMS model computes an enrichment factor for adsorbed chemicals based on changes of specific particle surface of the transported sediment caused by deposition in transport of the larger or more dense particles. Some management practices, such as land terracing and conservation tillage, are more effective in reducing transport of the coarser sediment and adsorbed chemicals are enriched in sediment compared to those in the residual soil and are increased as the mass of transported sediment decreases.[101] In Table 4, slight differences between pesticides C and D in percent transported by sediment may be related to differences in sediment enrichment factors later in the growing season which would affect transport of herbicide D, the more persistent pesticide.

Generalizations and comparisons above are based on 20-year averages. A simulation period at least this long is necessary to capture expected year-to-year variability in rainfall patterns to establish meaningful annual averages. Year-to-year variability in simulated herbicide runoff is illustrated for each of the herbicides and soils in Figure 10. Within the 20-year period runoff losses of herbicide A from the Houston soil ranged from 0.1% (4 years) to > 10% (2 years). The extent of loss of this nonpersistent herbicide is totally dependent on storm occurrence with respect to herbicide application. Runoff losses were relatively high when significant runoff-producing rainfall occurred on or near the application date. However, losses were minimal when runoff did not occur during this time period. The probability of herbicide loss increases with increasing herbicide half-life as shown by comparing losses of herbicide A and herbicide B (Figure 10, Houston soil). The effects of increasing half-life are not evident in comparing losses of A and B from the Tifton soil because leaching from the soil surface is the dominant process affecting these mobile compounds.

Losses of adsorbed herbicides C and D from both the Tifton and the Houston soils increased with increasing herbicide half-life, as expected. Losses of these compounds were dependent on occurrences of sediment-producing events during the time of persistence as determined by herbicide half-life. Therefore, the year-to-year variability in losses (Figure 10) is indicative of the variability in sediment-producing events during the period of pesticide persistence.

It is usually impractical to conduct experiments designed to continuously measure pesticide losses for more than 1 to 3 years. Observations in Figure 10 clearly demonstrate the necessity of long-term assessments to characterize expected year-to-year variability. Also, only a limited number of herbicides, soils, and management variables can be studied experimentally. Use of models such as CREAMS allows examination of many combinations of variables that otherwise would be impossible. The length of time simulated is limited only by the length of historical rainfall data available. In the absence of long-term weather records, climate generation models can be used to creat input data files of any length.[109]

D. Variability in Herbicide Runoff Prediction

In the preceding section, a 20-year simulation provided data for a rudimentary frequency analysis. The analysis could have been improved by using a longer simulation period. Longer term simulation, however, requires rainfall input data files that often are not available and stochastic rainfall generation models would be required. Another option in examining environmental risk in probabilistic terms is to use a stochastic pesticide state model, assuming that the state of the pesticide, the amount available for runoff at the time of a significant rainfall event, is a random variable and the primary variable affecting runoff losses. Mills and Leonard[70] developed such a model and demonstrated how probability density functions are generated specific to given pesticides, expected climatic patterns, and application methods and rates.

In developing the model, the amount of pesticide on foliage or soil available for runoff was assumed to be largely dependent upon the amount of pesticide applied, the percentage of the applied amount that is susceptible to runoff, the persistence or half-life of the pesticide, and the time from application to the occurrence of runoff-producing rainfall. The amount of pesticide applied, the pesticide half-life, and its susceptibility to runoff are not random variables, but are determined by pesticide chemistry, management decisions, and other conditions. Time from application to occurrence of significant rainfall is, however, subject to random variation. The probability distribution from the time of pesticide application to the occurrence of significant rainfall expressed in probability density function (PDF) form is

$$f_1(t_1) = \alpha \exp(-\alpha t_1)$$

where α is the reciprocal of average time between rainfall events, and t_1 is the time from pesticide application to rainfall occurrence. A first-order pesticide dissipation relationship can be assumed such that the amount of pesticide susceptible to runoff at time t is

$$c(t) = A \exp(-Bt)$$

where t is time since application of the pesticide, and A and B are parameters determined by the amount applied and the half-life of the pesticide. Combining the equation above with change of variables and simplification of the results yield

$$f_c(C) = \frac{\alpha}{B} \cdot A^{\frac{-\alpha}{B}} \cdot c^{\frac{\alpha}{B} - 1}$$

which is the derivation of a PDF for the amount of pesticide available for runoff, assuming that the last application is the major source of available pesticide. For multiple applications of relatively persistent compounds, the PDF can be generated for each application and convolved for the total amount available from all applications. In use of the equation, α is obtained from rainfall data at the location of interest. Evaluation of the PDF shows that for short half-life pesticides, the probability is high for low ranges of c and conversely, for long half-life pesticides, the probability is high for relatively large ranges of c. Given a range of c, deterministic runoff models can then be used to relate c to amounts of pesticide runoff expected at the given levels of probability. Continued development of techniques, such as coupling stochastic pesticide state models with stochastic models of rainfall characteristics and watershed state and deterministic models for event runoff of water, sediment, and chemicals and environmental damage functions, also demonstrated conceptually by Mills and Leonard,[70] can lead to realistic methods of environmental risk analysis associated with pesticide runoff.

VI. FUTURE RESEARCH IN HERBICIDE RUNOFF

The direction of future research on herbicide runoff may be dictated by unforeseen environmental concerns and, perhaps, radical changes in agricultural production practices and types of new herbicides available for weed control. However, several areas for productive research can be enumerated and are listed below, not necessarily in order of priority.

- Continued development of data bases to test concepts of fundamental runoff processes and comprehensive data bases to test and evaluate sophisticated simulation models. A definite need exists for runoff data on newer compounds and formulations, especially those developed for use in conservation tillage.
- Basic research on the nature of the effective soil surface and how this surface contributes chemicals to runoff, including effects of surface conditions such as roughness, crusting, etc. Effects of soil properties, rainfall, and runoff characteristics should also be considered.
- Research on herbicide fate on crop foliage, soil and crop residues, and how washoff characteristics affects runoff potential. Included in this area should be how the presence of grass sod and crop residues in intimate contact with the soil surface affects the effective runoff zone and the mixing and entrainment in runoff processes.
- Development of better submodels to describe herbicide persistence at the soil surface, particularly relating herbicide chemistry and susceptibility to degradation processes as mediated by soil and environmental variables and local microbial dynamics.
- Translation of findings from the above research into improved physically based models for continuous simulation.
- Continued development of methods for probabilistic analysis of environmental risks associated with herbicide runoff.
- Coupling of herbicide runoff and environmental risk models with models describing herbicide efficacy and crop yields for use with advanced generations of "smart" or artificial intelligence-generating computers. This future technology could allow real-time optimization of both production and environmental goals.

REFERENCES

1. **Nicholson, H. P., Grzenda, A. R., Lauer, G. J., Cox, W. S., and Teasley, J. I.,** Water pollution by insecticides in an agricultural river basin. I. Occurrence of insecticides in river and treated municipal water, *Limnol. Oceanogr.,* 9, 310, 1964.
2. **Nicholson, H. P., Grzenda, A. R., and Teasley, J. I.,** Water pollution by insecticides. A six and one-half year study of a watershed, Water Resources Center, University of California, Davis, *Proc. Symp. Agric. Waste Water,* 10, 132, 1966.
3. **Stewart, B. A., Woolhiser, D. A., Wischmeier, W. H., Caro, J. H., and Frere, M. H.,** in Control of Water Pollution from Cropland, Vol. 1, A manual for guideline development, ARS-H-5-1 and EPA-600/ 2-75-026a, U.S. Department of Agriculture and U.S. Environmental Protection Agency, Washington, D.C., 1975.
4. **Caro, J. H.,** Pesticides in agricultural runoff, in *Control of Water Pollution from Cropland,* Vol. 2, EPA-600/2-75-026b, U.S. Environmental Protection Agency and U.S. Department of Agriculture Rep. ARS-H-5-2, Washington, D.C., 1976, 91.
5. **Kemp, W. M., Means, J. C., Jones, T. W., and Stevenson, J. C.,** Herbicides in Chesapeake Bay and their effect on submerged aquatic vegetation, in Chesapeake Bay Program Technical Studies: A Synthesis, Part IV, U.S. Environmental Protection Agency, Washington, D.C., 1982, 503.
6. **Forney, D. R. and Davis, D. E.,** Effects of low concentrations of herbicides on submerged aquatic plants, *Weed Sci.,* 29, 677, 1981.
7. **Merkle, M. G. and Bovey, R. W.,** Movement of pesticides in surface water, in *Pesticides in Soil and Water,* Guenzi, W. D., Ed., Soil Science Society of America, Madison, Wisconsin, 1974, 99.
8. **Bailey, G. W., Swank, R. R., Jr., and Nicholson, II. P.,** Predicting pesticide runoff from agricultural land: A conceptual model, *J. Environ. Qual.,* 3, 95, 1974.
9. **Ahuja, L. R., Ross, J. D., and Lehman, O. R.,** A theoretical analysis of interflow of water through surface soil horizons with implications for movement of chemicals in field runoff, *Water Resour. Res.,* 17, 65, 1981.
10. **Ahuja, L. R. and Lehman, O. R.,** The extent and nature of rainfall-soil interaction in the release of soluble chemicals to runoff, *J. Environ. Qual.,* 12, 34, 1983.
11. **Meyer, D. L., Foster, G. R., and Romkens, M. J. M.,** Source of soil eroded by water from upland slopes, in Sediment Prediction Workshop Proceedings, ARS-S-40, U.S. Department of Agriculture, Washington, D.C., 1975, 177.
12. **Donigian, A. S., Beyerlein, D. C., Davis, H. H., and Crawford, N. H.,** Agricultural Runoff Management Model, Version II: Refinement and Testing, EPA-600/3-77-098, U.S. Environmental Protection Agency, Washington, D.C., 1977.
13. **Huff, D. D. and Kruger, P.,** The chemical and physical parameters in a hydrologic transport model for radioactive aerosols, *Proc. Int. Hydrol. Symp.,* 1, 128, 1967.
14. **Crawford, N. H. and Donigian, A. S.,** Pesticide Transport and Runoff Model for Agricultural Lands, EPA-660/2-74-013, U.S. Environmental Protection Agency, Washington, D.C., 1973.
15. **Bruce, R. R., Harper, L. A., Leonard, R. A., Snyder, W. M., and Thomas, A. W.,** A model for runoff of pesticides from small upland watersheds, *J. Environ. Qual.,* 4, 541, 1975.
16. **Steenhuis, T. S. and Walter, M. F.,** Closed form solution for pesticide loss in runoff water, *Trans. ASAE,* 23, 616, 1980.
17. **Haith, D. A. and Tubbs, L. J.,** Operational methods for analysis of agriculture nonpoint source pollution, Search: Agriculture 16, Cornell University, Ithaca, New York, 1981.
18. **Williams, J. R. and Hann, R. W.,** Optimal Operation of Large Agricultural Watersheds with Water Quality Constraints, Texas Water Resources Institute, Tech. Rep., 96, 1978.
19. **Frere, M. H., Onstad, C. A., and Holtan, H. N.,** ACTMO: An Agricultural Chemical Transport Model, *ARS-H-3,* Agricultural Research Service, U.S. Department of Agriculture, Washington, D.C., 1975.
20. **Sharpley, A. N., Ahuja, L. R., and Menzel, R. G.,** The release of soil phosphorus to runoff in relation to the kinetics of desorption, *J. Environ. Qual.,* 10, 386, 1981.
21. **Leonard, R. A., Langdale, G. W., and Fleming, W. G.,** Herbicide runoff from upland Piedmont watersheds — data and implications for modeling pesticide transport, *J. Environ. Qual.,* 8, 223, 1979.
22. **Smith, C. N., Leonard, R. A., Langdale, G. W., and Bailey, G. W.,** Transport of Agricultural Chemicals from Small Upland Piedmont Watersheds, U.S. Environmental Protection Agency Rep. EPA-600/3-78-056, U.S. Government Printing Office, Washington, D.C., 1978.
23. **Leonard, R. A. and Wauchope, R. D.,** The pesticide submodel, in CREAMS: A Field-Scale Model for Chemicals, Runoff, and Erosion from Agricultural Management Systems, Vol. 1, Conservation Research Rep. 26, Knisel, W. G., Ed., U.S. Department of Agriculture, Washington, D.C., 1980, 88.
24. **Haith, D. A.,** A mathematical model for estimating pesticide losses in runoff, *J. Environ. Qual.,* 9, 428, 1980.

25. **Leonard, R. A. and Nowlin, J. D.,** The pesticide submodel, in CREAMS: A Field-Scale Model for Chemicals, Runoff, and Erosion from Agricultural Management Systems, Vol. 2, Conservation Research Rep. 26, Knisel, W. G., Ed., U.S. Department of Agriculture, Washington, D.C., 1980, 304.
26. **Ingram, J. J.,** Chemical Transfer From a Saturated Soil into Overland Flow, Master's thesis, Colorado State University, Fort Collins, 1979.
27. **Ingram, J. J. and Woolhiser, D. A.,** Chemical transfer into overland flow, in *Proc. Symp. Watershed Management,* American Society of Civil Engineers, Boise, Idaho, 1980, 4.
28. **Wauchope, R. D.,** The pesticide content of surface water draining from agricultural fields: a review, *J. Environ. Qual.,* 7, 459, 1978.
29. **Mulkey, L. A. and Falco, J. W.,** Sedimentation and erosion control implications for water quality management, Proc. Natl., Symp. Soil Erosion and Sedimentation of Water, ASAE Paper No. 4, December 1977.
30. **Pionke, H. B.,** Form and sediment associations of nutrients (C. N. and P.) and pesticides, in *Fluvial Transport of Sediment-Associated Nutrients and Contaminants, Proc. of a Workshop,* Shear, H. and Watson, A. E. P., Eds., International Joint Commission, Kitchener, Ontario, 1977, 199.
31. **Rao, P. S. C., Nkedi-Kizza, P., Davidson, J. M., and Ou, L. T.,** Retention and transformation of pesticides in relation to nonpoint source pollution for croplands, in *Agricultural Management and Water Quality,* Schaller, F. W. and Bailey, G. W., Eds., Iowa State University Press, Ames, 1983, chap. 8.
32. **Green, R. E., Davidson, J. M., and Biggar, J. W.,** An assessment of methods for determining adsorption-desorption of organic chemicals, in *Agrochemicals in Soils,* Banin, A. and Kafkafi, U., Eds., Pergamon Press, New York 1980, 73.
33. **Rao, P. S. C. and Davidson, J. M., Eds.,** Retention and Transformation of Selected Pesticides and Phosphorous in Soil-Water Systems: A Critical Review, EPA 600/3-82-060, NTIS No. PB82, 256, 884, 1980.
34. **Novotny, V., Tran, H., Simsiman, G. V., and Chesters, G.,** Mathematical modeling of land runoff contaminated by phosphorus, *J. Water Pollut. Control Fed.,* 50, 101, 1978.
35. **Wauchope, R. D. and Sharpley, A. N.,** Nonpoint pollution of surface waters by chemicals: kinetic aspects of desorption of pollutants by runoff water, *Weed Sci. Soc. Am. Abstr.,* p. 47, 1984.
36. **Lorber, M. N. and Mulkey, L. A.,** An evaluation of three pesticide runoff loading models, *J. Environ. Qual.,* 11, 519, 1982.
37. **Nutter, W. L., Tkaus, T., Bush, P. B., and Neary, D. G.,** Simulation of herbicide concentrations in stormflow from forested watersheds, *Water Resour. Bull.,* 20, 851, 1984.
38. **Dale, J. E.,** Control of Johnson grass (Sorghum halepense) and volunteer corn *(Zea mays)* in soybeans *(Glycine max), Weed Sci.,* 29, 708, 1981.
39. **Banks, P. A. and Robinson, E. L.,** The influence of straw mulch on the soil reception and persistence of metribuzin, *Weed Sci.,* 30, 164, 1982.
40. **Nash, R. G.,** Dissipation rates of pesticides from soils, in CREAMS: A Field-Scale Model for Chemicals, Runoff, and Erosion from Agricultural Management Systems, Vol. 3, Conservation Research Rep. 26, Knisel, W. G., Ed., U.S. Department of Agriculture, Washington, D.C., 1980, 560.
41. **Willis, G. H., Spencer, W. F., McDowell, L. L.,** The Interception of applied pesticide by foliage and their persistence and washoff potential, supporting documentation, in CREAMS: A Field-Scale Model for Chemicals, Runoff, and Erosion from Agricultural Management Systems, Vol. 3, Conservation Research Rep. 26, Knisel W. G., Eds., U.S. Department of Agriculture, Washington, D.C., 1980, chap. 18.
42. **Willis, G. H. and McDowell, L. L.,** *Rev. Environ. Contam. Toxicol.,* 100, 1987, in press.
43. **Martin, C. D., Baker, J. L., Erbach, D. C., and Johnson, H. P.,** Washoff of herbicides applied to corn residue, *Trans. ASAE,* 21, 1164, 1978.
44. **Bovey, R. W., Burnett, E., Richardson, C., Merkle, M. G., Baur, J. R., and Knisel, W. G.,** Occurrence of 2,4,5-T and picloram in surface runoff in water of the Blacklands of Texas, *J. Environ. Qual.,* 3, 61, 1974.
45. **Trichell, D.W., Morton, H. L., and Merkle, M. G.,** Loss of herbicides in runoff water, *Weed Sci.,* 16, 447, 1968.
46. **McDowell, L. L., Willis, G. H., Southwick, L. M., and Smith, S.,** Methyl parathion and EPN washoff from cotton plants by simulated rainfall, *Environ. Sci. Tech.,* 18, 423, 1984.
47. **Rohde, W. A., Asmussen, L. E., Hauser, E. W., and Johnson, S. W.,** Concentrations of Ethoprop in the Soil and Runoff Water of a Small Agricultural Watershed, ARS-S-2, U.S. Department of Agriculture, Washington, D.C., 1979.
48. **Wauchope, R. D., Savage, K. E., and DeCoursey, D. G.,** Measurement of herbicides in runoff from agricultural areas, in *Research Methods in Weed Science,* Truelove, B., Ed., Southern Weed Scientific Society, 1977, chap. 5.
49. **Wauchope, R. D. and Leonard, R. A.,** Maximum pesticide concentrations in agricultural runoff: a semiempirical prediction formula, *J. Environ. Qual.,* 9, 665, 1980.

50. **Wauchope, R. D. and Leonard, R. A.,** Pesticide concentrations in agricultural runoff: available data and an approximation formula, supporting documentation, in CREAMS: A Field-Scale Model for Chemicals, Runoff, and Erosion from Agricultural Management Systems, Vol. 3, Conservation Research Rep. 26, Knisel, W. G., Eds., U.S. Department of Agriculture, Washington, D.C., 1980, 544.

51. **Baker, J. L. and Laflen, J. M.,** Runoff losses of surface applied herbicides as affected by wheel tracks and incorporation, *J. Environ. Qual.,* 8, 602, 1979.

52. **Baker, J. L. and Johnson, H. P.,** The effect of tillage systems on pesticides in runoff from small watersheds, *Trans. ASAE,* 22, 554, 1979.

53. **White, A. W., Jr., Harper, L. A., Leonard, R. A.,and Turnbull, J. W.,** Trifluralin volatilization losses from a soybean field, *J. Environ. Qual.,* 6, 105, 1977.

54. **Willis, G. H., McDowell, L. L., Parr, J. F., and Murphree C. E.,** Pesticide concentrations and yields in runoff and sediment from a Mississippi Delta watershed, in Proc. 3rd Interagency Sediment Conf., Denver, Colorado, March 1976, 53.

55. **Willis, G. H., McDowell, L. L., Murphree, C. E., Southwick, L. M., and Smith M.,** Pesticide concentrations and yields in runoff from silty soils in the lower Mississippi Valley, U.S.A., *J. Agric. Food Chem.,* 31, 1171, 1983.

56. **Scifres, C. J., Hahn, R. R., Diaz-Colon, J., and Merkle, M. G.,** Picloram persistence in semiarid rangeland soils and water, *Weed Sci.,* 19, 381, 1971.

57. **Barnett, A. P., Hauser, E. W., White, A. W., and Holladay, J. H.,** Loss of 2,4-D in washoff from cultivated fallow land, *Weeds,* 15, 133, 1967.

58. **Pionke, H. B. and Chesters, G.,** Pesticide-sediment-water interactions, *J. Environ. Qual.,* 2, 29, 1973.

59. **Leonard, R. A., Bailey, G. W., and Swank, R. R., Jr.,** Transport, detoxification, fate, and effects of pesticides in soil, water, and aquatic environments, *Land Application of Waste Material,* Soil Conservation Society of America, Ankeny, Iowa 1976, 48.

60. **Lichtenberg, J. J., Eichelberger, J. W., Dressman, R. C., and Longbottom, J. F.,** Pesticides in surface waters of the United States — a 5-year summary, 1964-1968, *Pestic. Monit. J.,* 4, 71, 1970.

61. **Schafer, M. L., Peeler, J. T., Gardner, W. S., and Campbell, J. E.,** Pesticides in drinking water, waters from Mississippi and Missouri rivers, *Environ. Sci. Technol.,* 3, 1261, 1969.

62. **Schulze, J. A., Manigold, D. B., and Andrews, F. L.,** Pesticides in selected western streams — 1968-1971, *Pestic. Monit. J.,* 7, 73, 1973.

63. **Glotfelty, D. E., Taylor, A. W., Isensee, A. R., Jersey, J., and Glenn, S.,** Atrazine and simazine movement to Wye River estuary, *J. Environ. Qual.,* 13, 115, 1984.

64. **Wu, T. L., Correll, D. L., Remenapp, H. E. H.,** Herbicide runoff from experimental watersheds, *J. Environ. Qual.,* 12, 3, 1983.

65. **Frank, R. and Sirons, G. J.,** Atrazine — its use in corn production and its loss to stream waters in southern Ontario, Canada 1975-1977, *Sci. Total Environ.,* 12, 223, 1979.

66. **Frank, R., Braun, H. E., Holdrinet, M. V., Sirons, G. J., and Ripley, B. D.,** Agriculture and water quality in the Canadian Great Lakes Basin. V. Pesticide use in 11 agricultural watersheds and presence in stream water 1975-1977, *J. Environ. Qual.,* 11, 497, 1982.

67. **Baker, D. B., Kruger, K. A., and Setzler, J. V.,** The concentrations and transport of pesticides in northwestern Ohio rivers — 1981, final report, *Gov. Rep. Announce. (U.S.),* 82, 1700, 1981.

68. **Schepers, J. S., Vavricka, E. J., Andersen, D. R., Wittmuss, H. D., and Schuman, G. E.,** Agricultural runoff during a drought period, *J. Water Pollut. Control Fed.,* 52, 711, 1980.

69. **Dudley, D. R. and Karr, J. R.,** Pesticide and PCB residues in the Black Creek Watershed, Allen County, Indiana — 1977-1978, *Pestic. Monit. J.,* 13, 155, 1980.

70. **Mills, W. C. and Leonard, R. A.,** Pesticide pollution probabilities, *Trans. ASAE,* 27, 1704, 1985.

71. **Mayeux, H. S., Jr., Richardson, C. W., Bovey, R. W., Burnett, E., Merkle, M. G., and Meyer, R. E.,** Dissipation of picloram in storm runoff, *J. Environ. Qual.,* 13, 44, 1984.

72. **Rohde, W. A., Asmussen, L. E., Hauser, E. W., Wauchope, R. D., and Allison, H. D.,** Trifluralin movement in runoff from a small agricultural watershed, *J. Environ. Qual.,* 9, 37, 1980.

73. **Novotny, V. and Chesters, G.,** *Handbook of Nonpoint Pollution: Sources and Management,* Environmental Engineering Ser., Van Nostrand Reinhold, New York, 1981.

74. **Nicks, A. D., Burt, J. B., Knisel, W. G., and Gander, G. A.,** Application of a field-scale water quality model for evaluation of watershed land treatment project, Paper no. 84-2633, in Proc. Am. Soc. Agricultural Engineers, New Orleans, Louisiana, December 1984.

75. **McDowell, L. L., Willis, G. H., Murphree, C. E., Southwick, L. M., and Smith, S.,** Toxaphene and sediment yields in runoff from a Mississippi U.S.A. Delta watershed, *J. Environ. Qual.,* 10, 120, 1981.

76. **Menzel, R. G.,** Enrichment ratios for water quality modeling, supporting documentation, in CREAMS: A Field-Scale Model for Chemical, Runoff, and Erosion from Agricultural Management Systems, Vol. 3, Conservation Research Rep. 26, Knisel, W. G., Ed., U.S. Department of Agriculture, Washington, D.C., 1980, 486.

77. **Ritter, W. F., Johnson, H. P., Lovely, W. G., and Molnau, M.,** Atrazine, propachlor and diazinon residues on small agricultural watersheds, *Environ. Sci. Technol.,* 8, 37, 1974.
78. **Baker, J. L., Laflen, J. M., and Johnson, H. P.,** Effect of tillage systems on runoff losses of pesticides, a rainfall simulation study, *Trans. ASAE,* 21, 886, 1978.
79. **Wauchope, R. D., McDowell, L. L., and Hagen, L. J.,** Environmental effects of limited tillage, in *Weed Control in Limited Tillage Systems,* Wiese, A. F., Ed., Weed Scientific Society of America Monogr., 1985, chap. 7.
80. **Baker, J. L., Laflen, J. M., and Hartwig, R. O.,** Effects of corn residue and herbicide placement on herbicide runoff losses, *Trans. ASAE,* 25, 340, 1982.
81. **Triplett, G. B., Jr., Conner, B. J., and Edwards, W. M.,** Transport of atrazine and simazine in runoff from conventional and no-tillage corn, *J. Environ. Qual.,* 7, 77, 1978.
82. **Edwards, W. M., Triplett, C. G., Jr., and Kramer, R. M.,** A watershed study of glyphosate transport in runoff, *J. Environ. Qual.,* 9, 661, 1980.
83. **Hall, J. K., Hartwig, N. L., and Hoffman, L. D.,** Cyanazine losses in runoff from no-tillage corn in "living" and dead mulches vs. unmulched conventional tillage (herbicide, *Zea mays*), *J. Environ. Qual.,* 13, 105, 1984.
84. **Langdale, G. W. and Leonard, R. A.,** Nutrient and sediment losses associated with conventional and reduced tillage, in *Agricultural Practices in Nutrient Cycling in Agricultural Ecosystems,* Lowrance, R. R., Todd, R. L., Asmussen, L. E., and Leonard, R. A., Eds., Special Publ. 23, University of Georgia, Athens, College of Agriculture Experiment Station, 1983, 457.
85. **Kramer, L. A.,** Seasonal runoff and soil loss from conventional and conservation tilled corn, Paper no. 84-2554, in Proc. Am. Soc. Agricultural Engineers, New Orleans, Louisiana, December 1984.
86. **Asmussen, L. E., White, A. W., Hauser, E. W.,and Sheridan, J. M.,** Reduction of 2,4-D load in surface runoff down a grassed waterway, *J. Environ. Qual.,* 6, 159, 1977.
87. **Crawford, N. H. and Linsley, R. D.,** The Synthesis of Continuous Streamflow Hydrographs on a Digital Computer, *Tech. Rep. No. 12,* Stanford University, Department of Civil Engineers, Stanford, California, 1962.
88. **Negev, M. A.,** Sediment Model on a Digital Computer, *Tech. Rep. No. 76,* Stanford University Department of Civil Engineers, Stanford, California, 1967.
89. **Donigian, A. S. and Crawford, N. H.,** Modeling Pesticides and Nutrients on Agricultural Lands, EPA-600/2-76-043, U.S. Environmental Protection Agency, Washington, D.C., 1976.
90. **Donigian, A. S., Imhoff, J. C., and Bicknell, B. K.,** in *Agricultural Management and Water Quality,* Schaller, F. W. and Bailey, G. W., Eds., Iowa State University Press, Ames, 1983.
91. **Holtan, H. W. and Lopez, N. C.,** USDAHL-70 Model of Watershed Hydrology, *Tech. Bull. 1435,* U.S. Department of Agriculture, Washington, D.C., 1971.
92. **Foster, G. R., Meyer, L. D., and Onstad, C. A.,** A runoff erosivity factor and variable slope length exponents for soil loss estimates, *Trans. ASAE,* 20, 683, 1977.
93. **Soil Conservation Service,** SCS National Engineering Handbook, Section 4, U.S. Department of Agriculture, Washington, D.C., 1972.
94. **Williams, J. R.,** Sediment yield predictions with universal equations using runoff energy factor, in Present and Prospective Technology for Predicting Sediment Yields and Sources, Agricultural Research Service, U.S. Department of Agriculture, Washington, D.C., 1975, 244.
95. **Steenhuis, T. S.,** Simulation of the action of soil and water conservation practices in controlling pesticides, in Effectiveness of Soil and Water Conservation Practices for Pollution Control, Haith, D. A. and Loehr, R. C.,Eds., EPA-600/3-79-106, U.S. Environmental Protection Agency, Washington, D.C., 1979, 106.
96. **Rao, P. S. C., Davidson, J. M., and Hammond, L. C.,** Estimation of nonreactive and reactive solute front locations in soils, in Residual Management By Land Disposal, Fuller, W. H., Ed., EPA-600/9-76-015, U.S. Environmental Protection Agency, Solid and Hazardous Water Research Division, Washington, D.C., 1976, 235.
97. **Knisel, W. G., Ed., CREAMS: A Field-Scale Model for Chemicals, Runoff, and Erosion from Agricultural Management Systems,** Conservation Research Rep. No. 26, U.S. Department of Agriculture, Washington, D.C., 1980.
98. **DeCoursey, D. G.,** ARS Small Watershed Model, ASAE Paper no. 82-2094, University of Wisconsin, Madison, Summer 1982.
99. **Computer Science Corporation,** Pesticide Runoff Simulation, User's Manual, Prepared for Office of Pesticides and Toxic Substances, U.S. Environmental Protection Agency, Falls Church, Virginia, 1980.
100. **Williams, J. R., Nicks, A. D., and Arnold, J. G.,** Simulation for water resources in rural basins, *ASCE J., Hydrol. Div.,* 1985, in press.
101. **DelVecchio, J. R. and Knisel, W. G.,** Application of field-scale nonpoint pollution model, *Proc. Am. Soc. Civil Engineers,* Irrigation and Drainage Specialty Conf., Orlando, Florida, 1982, 227.
102. **Haith, D. A. and Loehr, R. C., Eds.,** Effectiveness of Soil and Water Conservation Practices for Pollution Control, EPA-600/3-79-106 U.S. Environmental Protection Agency, Athens, GA., 1979.

103. **Williams, J. R. and LaSeur, W. V.,** Water yield model using SCS curve numbers, *ASCE J., Hydrol. Div.,* 102 (HY9), 1241, 1976.

104. **Green, W. A. and Ampt, G. A.,** Studies on soil physics. The flow of air and water through soils, *J. Agric. Sci.,* 4, 1, 1911.

105. **Foster, G. R., Lane, L. J., Nowlin, J. D., Laflen, J. M., and Young, R. A.,** A model to estimate sediment yield from field-sized areas: development of model, in CREAMS: A Field-Scale Model for Chemicals, Runoff, and Erosion from Agricultural Management Systems, Conservation Research Rep. 26, Knisel, W. G., Ed., U.S. Department of Agriculture, Washington, D.C., 1980, 36.

106. **Leonard, R. A. and Knisel, W. G.,** Model selection for nonpoint source pollution and resource conservation, in Proc. Int. Conf. Agriculture and Environment, Venice, Italy, 1984.

107. **Sheridan, J. M., Knisel, W. G., Woody, T. K., and Asmussen, L. E.,** Seasonal Variation in Rainfall and Rainfall-Deficient Periods in the Southern Coastal Plain and Flatwoods Regions of Georgia, Res. Bull. 243, University of Georgia, Athens, College of Agricultural Experimental Station, 1979.

108. **Wischmeier, W. H. and Smith, D. D.,** Predicting rainfall erosion losses — a guide to conservation planning, Agriculture Handbook No. 537, U.S. Department of Agriculture, Washington, D.C., 1978.

109. **Nicks, A. D.,** Stochastic generation of the occurrence, pattern, and location of maximum amount of daily rainfall, Misc. pub. 1275, Proc. Symp. Statistical Hydrology, U.S. Department of Agriculture Research Service, Washington, D.C, 1974, 154.

110. **White, A. W., Barnett, A. P., Wright, B. G., and Holladay, J. H.,** Atrazine losses from fallow land caused by runoff and erosion, *Environ. Sci. Technol.,* 1, 740, 1967.

111. **Bradley, J. R., Jr., Sheets, T. J., and Jackson, M. D.,** DDT and toxaphene movement in surface water from cotton plots, *J. Environ. Qual.,* 1, 102, 1972.

112. **Smith, S., Reagan, T. E., Flynn, J. L., and Willis, G. H.,** Azinphosmethyl and fenvalerate runoff loss from a sugarcane-insect IPM system, *J. Environ. Qual.,* 12, 534, 1983.

113. **Skaggs, R. W. and Khaleel, R.,** Infiltration, in *Hydrologic Modeling of Small Watersheds,* Haan, C. T., Johnson, II. P., and Brakensiek, D. L., Eds., American Society of Agricultural Engineers Monogr., St. Joseph, Michigan, 1982, 121.

114. **Gaynor, J. D. and Volk, V. V.,** Runoff losses of atrazine and terbutryne from unlimed and limed soil, *Environ. Sci. Tech.,* 15, 440, 1981.

115. **Rawls, W. J. and Brakensiek, D. L.,** Estimating soil water retention from soil properties, *Proc. Am. Soc. Agric. Eng.,* 108(IR2), 1982, 161.

116. **Knisel, W. G. and Baird, R. W.,** Runoff volume prediction from daily climatic data, *Water Resour. Res.,* 5, 84, 1969.

117. **Baldwin, F. L., Santelmann, P. W., and Davidson, J. M.,** Movement of fluometuron across and through the soil, *J. Environ. Qual.,* 4, 191, 1975.

118. **Hall, J. K.,** Erosional losses of *s*-triazine herbicides, *J. Environ. Qual.,* 3, 174, 1974.

119. **McDowell, L. L. and Grissinger, E. H.,** Erosion and water quality, in *Proc. 23rd Nat. Watershed Congr.,* Fletcher, D. A., Ed., National Association of Conservation Districts, Washington, D. C., 1976, 41.

120. **Dowler, C. A., Rohde, W. A., Fetzer, L. E., Scott, D. E., Sklany, T. E., and Swann, C. W.,** The effect of sprinkler irrigation on herbicide efficacy, distribution, and penetration in some Coastal Plain soils, *Georgia Agric. Exp. Stn. Res. Bull.,* 281, 1982.

121. **Spencer, W. F., Cliath, M. M., Blair, J., and LeMest, R. A.,** Transport of Pesticides from Irrigated Fields in Surface Runoff and Tile Drain Waters, Project Rep. EPA-IAG No. 78-D-X-0117, 1982.

122. **Hall, J. K., Pawlus, M., and Higgins, E. R.,** Losses of atrazine in runoff water and soil sediment, *J. Environ. Qual.,* 1, 172, 1972.

123. **Bailey, G. W., Swank, R. R., Jr., and Nicholson, H. P.,** Predicting pesticide runoff from agricultural land: a conceptual model, *J. Environ. Qual.,* 3, 95, 1974.

124. **Edwards, W. M.,** Agricultural chemical pollution as affected by reduced tillage systems, in Proc. No-Tillage Systems Symp., Ohio State University, Columbus, February 1972, 30.

125. **Wu, T. L.,** Dissipation of the herbicides atrazine and alachlor in a Maryland U.S.A. Corn field, *J. Environ. Qual.,* 9, 459, 1980.

126. **Rohde, W. A., Asmussen, L. E., Hauser, E. W., Hester, M. L., and Allison, H. D.,** Atrazine persistence in soil and transport in surface and subsurface runoff from plots in the Coastal Plain of the southern United States, *Agro-Ecosystems,* 7, 225, 1981.

127. **Sheets, T. J., Bradley, J. R., Jr., and Jackson, M. D.,** Contamination of surface and groundwater with pesticides applied to cotton, Water Resource Research Institute Rep. No. 60, University of North Carolina, Chapel Hill, 1972.

128. **Willis, G. J., Rogers, R. L., and Southwick, L. M.,** Losses of diuron, linuron, fenac, and trifluralin in surface drainage water, *J. Environ. Qual.,* 4, 399, 1975.

129. **Haas, R. H., Scifres, C. J., Merkle, M. G., Hahn, R. R., and Hoffman, G. O.,** Occurrence and persistence of picloram in grassland water sources, *Weed Res.,* 11, 54, 1971.

130. **Glass, B. L. and Edwards, W. M.,** Picloram in lysimeter runoff and percolation water, *Bull. Environ. Contam. Toxicol.,* 11, 190, 1974.
131. **Baur, J. R., Bovey, R. W., and Merkle, M. G.,** Concentration of picloram in runoff water, *Weed Sci.,* 20, 309, 1972.
132. **Scifres, C. J., McCall, H. G., Maxey, R., and Tai, H.,** Residual properties of 2,4,5-T and picloram in sandy rangeland soils, *J. Environ. Qual.,* 6, 36, 1977.
133. **Johnsen, T. N., Jr.,** Picloram in water and soil from a semiarid Pinyon-Juniper watershed, *J. Environ. Qual.,* 9, 601, 1980.
134. **White, A. W., Asmussen, L. E., Hauser, E. W., and Turnbull, J. W.,** Loss of 2,4-D in runoff from plots receiving simulated rainfall and from a small agricultural watershed, *J. Environ. Qual.,* 5, 487, 1976.
135. **Norris, L. A., Montgomery, M. L., Warren, L. E., and Mosher, W. D.,** Brush control with herbicides on hill pasture sites in southern Oregon, *J. Range Manage.,* 35, 75, 1982.
136. **Nicholaichuk, W. and Grover, R.,** Loss of fall applied 2,4-D in spring run-off from a small agricultural watershed, *J. Environ. Qual.,* 12, 412, 1983.

Chapter 4

EVAPORATION FROM SOILS AND CROPS

Alan W. Taylor and Dwight E. Glotfelty

TABLE OF CONTENTS

I. GENERAL INTRODUCTION

Herbicides begin to disappear from the target area immediately after application by either the action of air or water, or by chemical degradation with the formation of new compounds. Both processes impose limits on the effective life of the herbicide because they reduce the length of time that the residues are concentrated enough to inhibit the growth of the target species.

The rate of chemical degradation reflects the stability of the compound when exposed to the chemical or biochemical environment existing on the soil or plant surface. While degradation reactions may be sensitive to environmental factors such as temperature, light intensity, or humidity, they can be predicted with considerable certainty from data obtained under controlled conditions in the laboratory or greenhouse. When such data are, however, applied to the prediction of overall disappearance rates in the field, it is often evident that factors other than chemical degradation must be taken into account, and may indeed sometimes be dominant.

In considering these physical losses it is very difficult to generalize about relative amounts of herbicides that may be carried out of the target area by air or water. Leaching and runoff are totally dependent upon rainfall and weather patterns and may be difficult to predict; except in very special circumstances, the total seasonal losses by this pathway rarely appear to exceed 5 or 10% of the total applied.[1] In contrast, under unfavorable conditions, losses of residues by volatilization may sometimes approach 80 to 90% within a few days, although such rates are also sensitive to weather and microclimatic conditions.

The importance of volatilization as a mechanism controlling the persistence of residues of agrochemicals in the field is not confined to herbicides and is also a prominent factor controlling the effectiveness of many insecticides. It is indeed notable that much of the early research on the volatilization of agrochemicals was done with insecticides. The basic reason for this was that many of the insecticides were chemically very stable and could be transported very great distances in the atmosphere, and then reconcentrated by bioaccumulation in nontarget organisms, making the volatility of these compounds as a matter of primary environmental interest. A second and more technical reason was that their stability gave them a conservative character so that their chemical degradation rates could be discounted in measuring disappearance times. Where suitable data on herbicides is not available, data drawn from this earlier work on insecticides will be used for this chapter. From a fundamental point of view, no distinctions need be drawn between the two classes of compounds in terms of environmental behavior.

The factors that control the rate of volatilization are clearly of prime interest in understanding the role it plays in the loss of efficiency and the environmental distribution and impact of herbicides. The primary emphasis of this chapter will therefore be directed to a

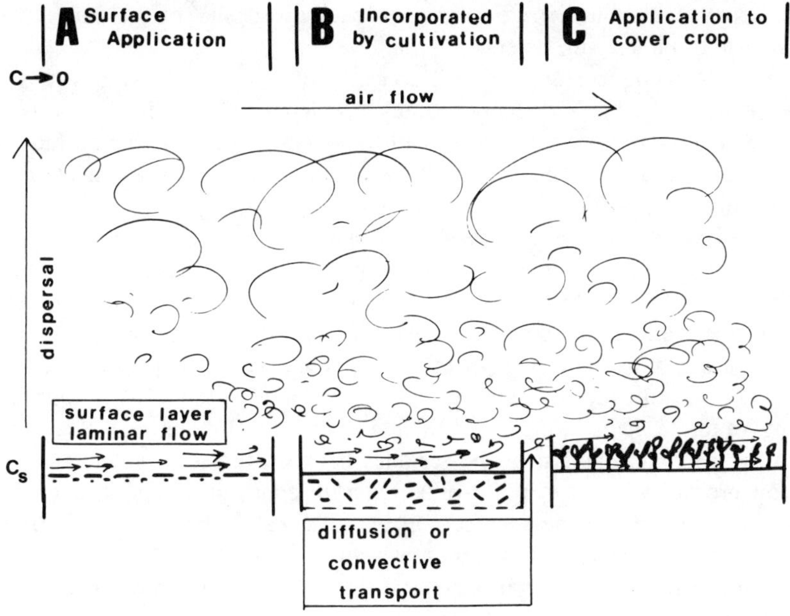

FIGURE 1. Physical distribution of herbicide residues in soil under various field management systems and the resulting pathways of dispersal to the air.

discussion of the underlying physics and chemistry of the processes involved. This discussion will be essentially confined to processes acting at or close to the surfaces of the soil or plant leaves and in the atmosphere a few meters above them. Questions concerning long-distance transport of herbicides in the atmosphere, including their chemical and photo-chemical stability in air and their rates of redeposition in rain or fog, touch upon more general aspects of their environmental chemistry. These will not be discussed here in detail. Problems of herbicide injection into the atmosphere by evaporation or drift of spray droplets will also be considered elsewhere.

II. BASIC PHYSICAL PROCESSES

Two physicochemical processes control the dispersal of herbicides into the environment. These are the evaporation of herbicide molecules into the air from liquids, solutions, or solids present either as formulations or pure active ingredients adsorbed on soil or plant surfaces, and second, the dispersion and dilution of the resulting vapor through the overlying atmosphere. The ways in which these operate and the factors involved in each are illustrated in Figure 1.

Herbicides are usually applied to the surfaces of bare soil (Figure 1A), which may sometimes be partly covered by plant residues from the previous crop or by turf or low grass cover crops (Figure 1B). Since these residues disperse directly into the overlying air, one of the principal factors controlling their rate of loss will be the rate at which the surface residues evaporate. Such residues may not, however, be present as a simple thin layer of material at the soil/air interface. Where herbicides are applied to dry soils as liquid sprays of solutions or emulsions, some fraction of the residues will be drawn into the outer surface layer of soil by capillary action or diffusion. Initially, the volatilization of the herbicide applied in this way will depend upon the vapor pressure of the residues exposed on the surface. This will be that of the active ingredient itself or the lower values to which this is reduced by adsorption on the soil material. After the outer surface layer is depleted, the rate may become dependent upon the rate of single "back-diffusion" of the material adsorbed into the immediately underlying layers.

It is important to recognize that the soil concentrations of such residues may be quite high: if a herbicide applied at 1 kg/ha penetrates to a depth of 2 mm into a soil with a bulk density of 1.25, the average concentration will be about 40 μg/g. If the surface coverage is only 30%, with residues confined to the areas of drop impact, local concentrations may be about three times this value.

Where residues are not allowed to remain on the soil surface, but are incorporated to some depth by cultivation, as is often the practice with more volatile preemergence herbicides, such as trifluralin, much smaller amounts of residues will remain on the surface and the average soil concentration will be much lower (Figure 1C). Incorporation of a 1 kg/ha application to a depth of 7.5 cm (= 3 in.) will reduce the average soil concentration to 1.0 μg/g or less. The volatilization of such residues will clearly be dependent upon the rate at which they move to the open surface by diffusion or by bulk flow in the soil water; the role of adsorption in the redistribution of the residues between the soil, water, and air phases within the soil profile will be of major importance in controlling this process. It should, however, be remembered that the average concentration calculated from the ratio of application rate/cultivation depth does not represent actual soil values. Cultivation by field implements is a very ineffective mixing process, and the actual residues may remain as "grains" or "hot spots" of high concentration scattered through the soil body. Movement of these residues may, therefore (at least initially), involve diffusive movement through short distances with high concentration gradients out of localized volumes of "saturated" soil into surrounding regions of low residue content and high adsorption capacity.

The flow of air over plant and soil surfaces is invariably turbulent except in a very shallow layer of about 1 mm in depth, where the flow may be regarded as laminar and controlled by viscosity.[2] The two regions of laminar and turbulent flow may be distinguished as those where the dispersal of pesticide vapor can be described in terms of molecular diffusion coefficients (the surface laminar layer of streamline flow), and the lower layers of the overlying atmosphere where flow is turbulent and dispersal must be described in terms of the "eddy diffusion" coefficient (Figure 1). The depth of the turbulent zone, often described as the "atmospheric boundary layer", is several orders of magnitude greater than that of the laminar flow layer, being measured in meters rather than the centimeter or millimeter depths of the latter. The difference between the two diffusion coefficients is very large, so that except when the atmosphere is exceptionally stable, dispersal of the pesticide vapor in the turbulent zone outside the laminar layer is relatively very rapid. It may therefore be expected that the rate-limiting steps will be found within the layer of laminar flow where molecular processes are dominant.

The upward flux away from the surface through the laminar layer of depth m may be described by the equation below:

$$F = D_m(d_s - d)/m \qquad (1)$$

where F is flux per unit area, D_m is the molecular diffusion coefficient, d_s is the concentration of vapor (vapor density) at the surface, and d is the vapor density at the top of the laminar layer. The significance of this equation has been discussed in detail by Hartley and Graham-Bryce,[3] who point out that the parameter m is not directly measurable and will also vary with the wind speed and the character of the surface. Although estimates of its value are important in some cases, as in studies of the transport of water and carbon dioxide, where it contributes resistance to the flow of these into or out of leaf surfaces, in the present case, the other variables in Equation 1 are more important, emphasizing the role of the diffusion coefficient and the concentration gradient $(d_s - d)$ in controlling the pesticide flux.

The upward flux or dispersal of pesticide through the turbulent boundary layer is described by the following equation:

$$\uparrow P = K_p(dp/dz) \tag{2}$$

where $\uparrow P$ is the vertical flux, dp/dz is the gradient of herbicide vapor density, and K_p is the diffusivity coefficient for the dispersal of the vapor in turbulent air (the "eddy diffusivity coefficient") both at the height z above the surface. This equation is of fundamental importance in the interpretation of experimental measurements of herbicide fluxes through the turbulent boundary layer. This topic will be considered in more detail below.

The process of herbicide dispersal can be described in fundamentally the same terms for each of the environmental phases in which it takes place. In each case, as in Equations 1 and 2, the flux rate ($\uparrow F$ or $\uparrow P$) is related to a concentration gradient ΔC existing over a distance of ΔZ by Fick's law of diffusion:

$$\uparrow F = -D\Delta C/\Delta Z$$

where D is the diffusivity coefficient for the herbicide within the particular phase. When, as in the dispersal of herbicides from within the soil profile into the overlying air, we must consider movement through a series of phases, the equations are expressed in different units for each phase (e.g., concentrations in $\mu g/g$ for soil, vapor pressures [pascals] or vapor densities [$\mu g/L$] in air), and the diffusivity coefficients are also treated separately. The dispersal process can, however, be regarded as a single unified whole if two changes are made in the approach. The first is to rewrite the flux equation as $F = K\Delta C$, where $K = D/\Delta Z$, thus transforming the diffusivity coefficient into a mass transfer coefficient. The reciprocal of this transfer coefficient is a resistance term which is then analogous to those describing conductive heat or electrical flow as in Ohm's law. This has the great advantage, as pointed out by Mackay,[4] that if the concentration or potential terms are converted to suitable units, the equations for all the phases are additive, and the overall flux can be described in terms of a single potential gradient over the whole path. This can be readily achieved by the use of the fugacity concept, originally postulated by Lewis[5] and reintroduced by Mackay[6] as a fundamental principle in environmental chemistry. In this approach fugacity is defined as the "escaping tendency" of a component (the herbicide), and has the dimensions of pressure. In using it to interpret movements of herbicides through the environment, it is necessary to relate the vapor pressure (or vapor density) in the gas phase to the concentration in solution by the use of the Henry's law relation, $p = k_h \cdot c^*$, and to use adsorption isotherms stated in terms of equilibrium vapor pressures for characterization of the residues on the solid phase. Since, as will be described below, this is particularly convenient in experimental measurements with volatile herbicides, the approach has many advantages and, as shown by Mackay,[6] provides a powerful tool in the prediction of the movements of residues through the various phases of the environment. For the purposes of the present discussion, it will be used as a useful conceptual and simplifying approach in which herbicide dispersal is regarded as a single flow path through a series of barriers, each with a characteristic, but additive resistance. Since, however, much of existing literature is presented in terms of diffusivities, both approaches will be used interchangeably in the present discussion.

A. Equilibrium Vapor Pressures

Table 1 contains a number of values of measured vapor pressures of commonly used herbicides and insecticides selected from the literature to illustrate the range and scale of the values found.[7-9] The range is clearly very wide, extending over about six orders of

---

* Written here in terms of pressure, not vapor density.

Table 1
SATURATION VAPOR PRESSURES OF PESTICIDES

Compound	Molecular weight (g)	Temperature (°C)	Vapor pressure (mPa)	Vapor density (g/ℓ)	Ref.
Alachlor	270	25	2.9	3.2 (−1)[a]	9
Atrazine	216	20	4.0 (−2)	3.5 (−3)	7,9
Bromacil	261	25	2.9 (−2)	3.0 (−3)	8
Carbofuran	221	25	1.1	1.0 (−1)	8
DDT (general)	354	25	4.5 (−2)	6.0 (−3)	8
(pp)			2.5 (−2)	3.8 (−3)	7
Dicamba	221	20	4.9 (+2)	4.5 (+1)	7
Diazinon	304	20	1.9 (+1)	2.3	7
Dieldrin	381	25	6.8 (−1)	1.0 (−1)	8
Diuron	233	25	2.1 (−2)	2.0 (−3)	8
EPTC	189	25	2.8 (+3)	2.2 (+2)	8
Fenitrothion	277	20	8.0 (−1)	9.1 (−2)	7
Heptachlor	373	20	2.1 (+1)	3.3	7
Lindane	291	25	8.6	1.0	8
Linuron	249	20	1.2	1.2 (−1)	7
Malathion	330	20	1.3	1.8 (−1)	7
Metolachlor	284	20	1.7	2.0 (−1)	7,9
Monuron	199	25	2.3 (−2)	2.0 (−3)	8
Parathion	291	25	1.3	1.5 (−1)	8
Picloram	241	20	7.3 (−4)	7.3 (−5)	7
Simazine	202	25	2.0 (−3)	1.7 (−4)	8
Triallate	305	20	2.6 (+1)	3.2	8
Trifluralin	335	20	1.5 (+1)	2.0	8

[a]   Figures in parentheses represent power of 10 factors.

magnitude from 2800 mPa for EPTC to 0.002 mPA for simazine. Since the list is not comprehensive, other compounds with values outside this range may be found. Despite this wide range, the maximum values are very low in comparison with more familiar volatile organic chemicals, such as alcohols or ethers, which have vapor pressures close to one atmosphere at normal or slightly elevated temperatures. In making this comparison it may be noted that the pressure of a standard atmosphere (760 mm Hg) is equal to 1×10^5 Pa, or about 4×10^4 or 1×10^5 times that of EPTC. While the tabulated values would perhaps seem to suggest that the volatilization rates of the compounds would be very low, it must be remembered that the background vapor pressures of pesticides in the atmosphere are essentially zero, and exposed residues will therefore evaporate as if they were exposed to a vacuum. This effect is a direct consequence of Dalton's law, which states that the vapor pressure of a single gas is independent of those of any others that may be mixed with it. The absence of any natural background concentration places pesticide evaporation in singular contrast to those of other atmospheric components where a natural background concentration exists. The most striking example is that of water vapor, which has a vapor pressure of about 0.023 atmospheres or 2300 Pa in saturated air at 20°C. The limit which this places on the evaporation of water is a familiar feature of humid climates; as a true comparison, the evaporation of pesticides at the surface of the earth should be compared with that of water under the desiccated conditions on the surface of the moon or the planet Mars.

In some instances, the data in Table 1 include more than one value for a particular compound. These reflect differences in the published values due to uncertainties arising from different methods used in their determination. The measurement of the "true" values of such low vapor pressures presents considerable experimental difficulties. These methods and the uncertainties they entail have been reviewed in detail by Plimmer,[10] but will be also briefly described here since, in some cases, the same techniques have been used to measure the vapor pressures of residues in soil and water systems as well as those of the pure compounds.

The basic methods may be grouped into several distinct categories. In the simplest of these, a measurement of vapor pressure is made directly upon vapor in an equilibrium condition. This is most satisfactory where the vapor pressure is of the order of 130 Pa ($=$ 1 mm Hg) or more. This is much greater than that of pesticides at room temperatures, so that elevated temperatures must be used, and the results extrapolated to the normal range by equations based upon the integrated Clapeyron-Clausius equation expressed in the form

$$\log P = A - B/(t°C + 273.17)$$

A method of this type, in which a differential thermal analyzer was used to determine the boiling point of a liquid at an imposed pressure, was described by Hamaker and Kerlinger.[11] Data obtained from such high temperature methods have the disadvantage of uncertainties due to errors in both the experimental measurement and in the extrapolation. Where the extrapolation crosses the temperature where there is a phase change, the error may be considerable.

Effusion methods depend upon measurement of the rate of escape of vapor through a small orifice into a vacuum. The measurement may be either the weight loss of the container, or of the torque imposed on a fiber supporting the cell, or by collection of the escaping vapor in a liquid nitrogen cold trap. Several of these methods have been described in detail by Hamaker and Kerlinger.[11]

The gas saturation method described by Spencer and Cliath[12] has been used to measure vapor pressures of pesticides both as pure compounds and as residues in soils. This is a direct method in which the amount of vapor in a given volume (the vapor density) is determined by analysis of a volumetric sample taken at a suitable controlled temperature. The vapor pressure is then calculated from the classical equation:

$$P = (W/V)(RT/M) \tag{3}$$

where W is the weight of pesticide vapor, V is the sample volume, R is the gas constant in appropriate units, T is the absolute temperature and M is the molecular weight of the pesticide. Since the chromatographic methods of analysis permit accurate determination of nano- and picogram quantities of pesticides, measurements can be made in the 10 to 40°C range and no extrapolations are necessary; the principal assumption lies in the calculation of the vapor pressure from the vapor density, where use of the molecular weight implies that the vapor is completely dissociated.

Vapor pressures may also be calculated from gas/liquid chromatographic retention times. The accuracy of this method depends upon the assumption that the retention by the column depends only on the vapor pressure, so that comparison of the retention times of a series of compounds with that of one or more standard substances of known properties permits calculation of their relative vapor pressures. Successful application of the technique requires a suitable choice of both column materials and of standard substances suitable to the polarity and chemical character of the particular pesticides. The advantage of the method is the speed at which data can be obtained and its applicability to mixtures and impure samples. The principal disadvantage is, as with direct measurements made at elevated temperatures, the need to extrapolate the results to environmental temperatures. With the gas chromatographic method, one obtains the liquid-phase vapor pressures at all temperatures, even where these are below the melting point of the pesticide. Specific interactions with the gas chromatograph liquid phase may also cause problems.

Of these four general techniques, the best and most consistent results appear to be obtained by the gas saturation and gas/liquid chromatographic methods. Where critical values are sought from the literature, those obtained by these means should be preferred, provided that they are supported by an adequate presentation of the original data and description of the experimental conditions. Unfortunately, this is far from always the case for many published values. Few critical reviews exist, and where data are required for critical purposes, the results should be traced, if possible, to their origin. An example of the discrepancies that may be found has been discussed by Plimmer[10] for data on several esters of 2,4 dichloro-phenoxy-acetic acid.

B. Adsorption, Solubility, and Partition Effects

The value of d_s in Equation 1 is the vapor density established at the inner surface of the laminar flow layer of air moving over a surface. This will rarely be the saturation vapor density of the pure pesticide. The pesticide is always adsorbed on soil or plant material, or dissolved in water that is present on the surface or occluded within soil (or leaf) pores. As the residues equilibrate with the soil and plant surfaces after application, the vapor pressure in the overlying air is reduced below d_o, the equilibrium value of the pure compound, by adsorption reactions with the soil and plant surfaces. The degree to which the vapor pressure is reduced by soil adsorption depends upon a number of factors which include the soil water content, the soil texture, and organic matter content. Although there is much less information available about the effects of adsorption on plant surfaces, sorption and permeation into cuticular waxes and the epidermal layer of plant leaves are certainly important in reducing the pesticide vapor pressure; the complexities of herbicide penetration in regard to their effectiveness has been discussed in detail by Hartley and Graham-Bryce.[13]

Although there are a number of different patterns in the physicochemical behavior of pesticides in soils, as may be expected from the variety of types of chemicals used, many share a number of common features that are of basic importance in controlling their partition between the air, water, and solid phases of the soil, and hence, their rates of volatilization. Since the bulk of the residues is deposited on the soil solids, the most important features are those which govern the approach to equilibrium conditions within the soil system (uniform fugacity or chemical potential in all three phases when the system is closed). This implies desorption from the solid surface into the soil solution and air within the soil pores or the overlying laminar layer. As noted by Spencer,[14,15] the partitioning between the solid phase and the solution usually follows the Freundlich equation and the vapor density of the pesticide in the air, in turn, reflects the solution concentration according to Henry's law, defined by the equation below:

$$d = k_h \cdot c \tag{4}$$

where d is the vapor density and c is the solution concentration, both in the same units (usually milligrams or micrograms per liter). Values for the Henry's law coefficients, and water solubilities of the pesticides listed in Table 1 are presented in Table 2.

Measurements of pesticide concentrations in air equilibrated with treated soils can be made very conveniently under controlled conditions in the laboratory using the gas saturation technique,[12] in contrast to measurements of concentrations in the soil solution, which are less reliable, much more laborious, and require destruction of the sample. The vapor pressure measurements, therefore, offer a direct experimental route to the evaluation of adsorption effects on volatility. Data obtained in this way for several major pesticides have been published by Spencer and colleagues.[12,16-19]

Table 2
SOLUBILITY AND HENRY'S LAW
COEFFICIENTS

Compound	Solubility (mg/ℓ)	Coefficient $k_H = d_o/c_o$	Ref.
Alachlor	2.4 (+2)[a]	1.3 (−6)	7,9
Atrazine	3.2 (+1)	2.5 (−7)	8,9
Bromacil	8.2 (+2)	3.7 (−9)	8,9
Carbofuran	3.2 (+2)	3.1 (−7)	8
Chlorpropham	8.9 (+1)	1.2 (−6)	7,9
DDT (general)	3.0 (−3)	2.0 (−3)	8
(pp)	1.2 (−3)	3.3 (−3)	7
Diazinon	4.0 (+1)	5.8 (−5)	7
Dicamba	4.5 (+3)	1.0 (−5)	7
Dieldrin	1.5 (+2)	6.7 (−4)	8
Diuron	3.7 (+1)	5.4 (−8)	8
EPTC	3.7 (+2)	5.9 (−4)	8,9
Heptachlor	5.6 (−3)	5.9 (−2)	7
Lindane	7.5	1.3 (−4)	8
Linuron	7.5 (+1)	1.6 (−6)	7,9
Malathion	1.5 (+2)	1.3 (−6)	7,9
Metolachlor	5.3 (+2)	3.7 (−7)	7,9
Monuron	2.6 (+2)	7.7 (−9)	8
Parathion	2.4 (+1)	6.2 (−6)	8
Picloram	4.3 (+2)	1.7 (−10)	7
Simazine	5.0	3.4 (−8)	8
Triallate	4.0	8.0 (−4)	7,8
Trifluralin	3.0 (−1)	6.6 (−3)	8,9

Note: All data at 25°C.

[a] Figures in parentheses represent power of 10 factors.

1. Adsorption

Vapor phase desorption isotherms for trifluralin and dieldrin on Gila silt loam, obtained by Spencer et al.,[16,19] are presented in Figures 2A and B. The vapor densities of both compounds increase with increasing residue concentrations, reaching values corresponding to those of the pure compounds at soil concentrations of 73 µg/g for trifluralin and 25 µg/g for dieldrin. Other data, for lindane, also obtained by Spencer,[17] are presented in Figure 3. Here, the vapor densities are expressed in terms of the relative vapor density, d/d_o, rather than simple concentrations. These data show that the vapor densities supported by the residues in Gila soil approach the pure compound values at concentrations ranging from 50 to over 60 µg/g as the temperature increases above 20°C.

These results are believed to be typical of many other soils. In order to relate them to field conditions, we may assume that if a spray of herbicide applied at a rate of 12.5 kg/ha to the surface of a bare soil without cultivation remains within the top 1-mm layer of soil (bulk density 1.25), the local concentration may be expected to be about 75 to 100 µg/g. The data of Figures 2 and 3 suggest that at these concentrations the residues will support vapor densities close to those of the parent compounds, provided that the soil moisture levels are above about 4% or over. If, however, the herbicide is incorporated by cultivation to a greater depth, say 75 mm, the average soil concentration will be reduced to between 1 or 1.25 µg/g. The data of Figures 2 and 3 then suggest that the relative vapor densities will be reduced to below 10% or less. Equation 1 then suggests that there will then be a considerable reduction in the volatilization rate. However, as will be seen from the following discussion, other factors, including diffusion characteristics and moisture flow, also greatly

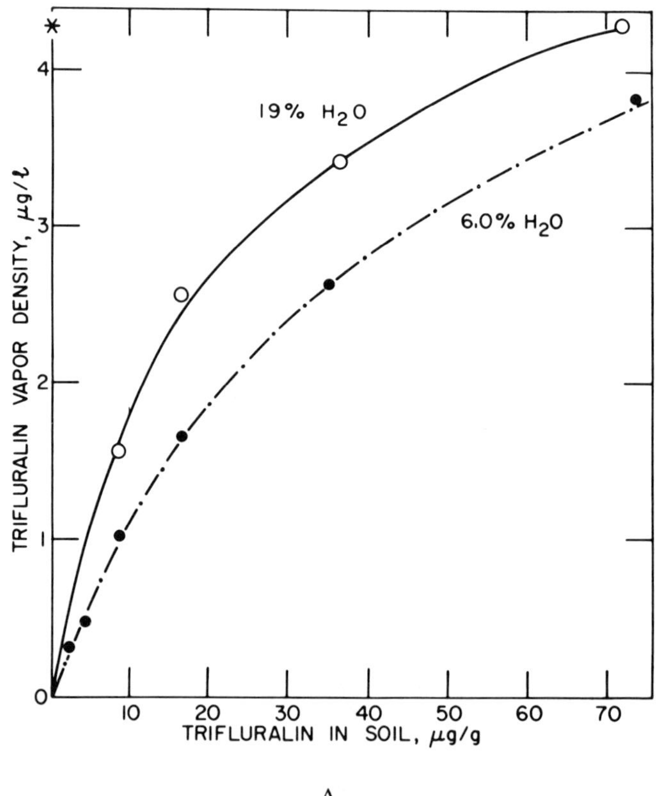

FIGURE 2.   Vapor densities of trifluralin and dieldrin residues as functions of concentration in Gila silt loam soil. (A) Trifluralin vapor density at 30°C and 6 and 19% soil water content. (* indicates vapor density of trifluralin without soil) and (B) dieldrin vapor density at 20, 30, and 40°C and soil water contents above 10%. (From (A) Spencer, W. F. and Claith, M. M., *J. Agric. Food Chem.*, 22, 987, 1974; and (B) Spencer, W. F., Claith, M. M., and Farmer, W. J., *Soil Sci. Soc. Am. Proc.*, 33, 509, 1969. Both with permission.)

affect the rates at which incorporated residues move to the open soil surface where they volatilize.

2. Temperature Effects on Adsorption

The data in Figures 2 and 3 indicate that the effect of temperature on the vapor densities supported by surface residues of high concentration will be similar to the effect of temperature on the vapor densities of the pure compounds. This is equivalent to saying that when the residue concentration exceeds the adsorption capacity in a moist soil, the heat of vaporization of the residues is the same as that of the pure compound.[12] Where residue concentrations are below the saturation level and the relative vapor densities (d/d_o) are less than 1.0, the temperature relations are more complex. In a detailed analysis of the lindane system, shown in Figure 4, Spencer and Cliath[17] point out that for a given soil residue concentration below the saturation level, the vapor density increases with temperature more slowly than that of the pure compound; this effect is due to a combination of solubility relationships (Henry's law, Equation 4) and energy contributions to the adsorption process. One result of these temperature functions is that, as shown in Figure 3, the residue concentration corresponding to the "saturation capacity", where $d/d_o = 1.0$, increases with temperature.

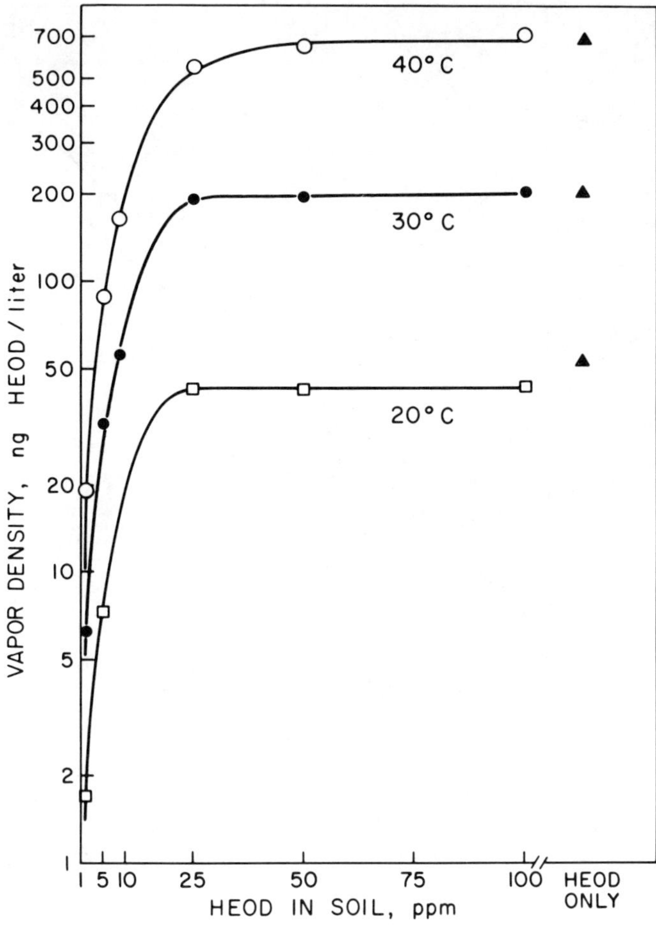

FIGURE 2B.

3. Henry's Law Distribution

The validity of the Henry's law relationship in describing the partitioning of the pesticide between the air and water phases of the soil is demonstrated by the curve plotted in Figure 4, which shows that all the lindane data for a particular temperature (30°C) fall on the same curve, and are independent of the soil moisture content when plotted in terms of d/d_o and the corresponding solubility function c/c_o. The values of c/c_o, the lindane concentrations in the soil solution, were obtained by analysis of the centrifugate of a 1:5 soil:water suspension. This result also validates the use of vapor measurements in the determination of adsorption (or desorption) isotherms of pesticides in soils, although the curves obtained include the effects of changes in the Henry's law coefficients (K_h) with concentration and temperature. A complete calculation of the distribution of residues between soil, water, and air would require independent measurement of these changes, but for the interpretation of the effects of adsorption on the vapor densities in equilibrium with the soil residues, this is not necessary because if Henry's law is obeyed, $d/d_o = c/c_o$ and the solution and adsorption effects may then be treated as one.

4. Effects of Soil Moisture Content

Retardation of the volatilization or disappearance of herbicides from dry soil has been noted by a number of investigators.[20-24] The underlying cause of this retardation, which is

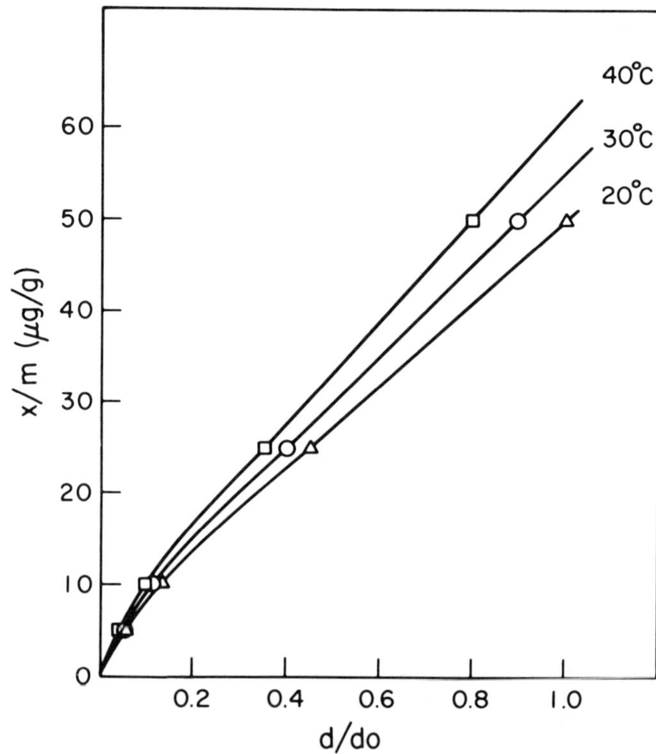

FIGURE 3.   Equilibrium relative vapor density of lindane residues as a function of concentration in Gila silt loam soil at 3.94% soil water content. (From Spencer, W. F. and Claith, M. M., *Soil Sci. Soc. Am. Proc.*, 34, 574, 1970. With permission.)

due to the increased adsorption of herbicides in dry soil, has been discussed in detail by Spencer[14] on the basis of the data presented in Figures 5A and B. These results show that even where the concentrations of trifluralin and dieldrin in Gila soil are high enough to support vapor densities close to those of the pure compounds when the soil is sufficiently moist, reduction of the soil water content to less than about 2.8% results in a large reduction in the equilibrium vapor densities; at water contents of 2% or less, the relative vapor densities are below 10%, and at 1.6% water content, the dieldrin vapor density approached the lower levels of detection. Similar data have been presented for DDT and DDE isomers[18] and for lindane.[14,17] A water content of 2.8% in the Gila soil corresponds to a monomolecular water layer on the adsorption surfaces. It is thus evident that when the moisture content is reduced to this level, adsorption sites are exposed on which the pesticide molecules are more strongly bound so that their fugacity (and hence, their vapor density) is greatly reduced. The effect is reversible in response to changes in water content for both trifluralin and dieldrin adsorption.[16,19]

In practical terms, it may be noted that in the Gila soil a moisture content of 3.9% is in equilibrium with air at a relative humidity of 94%. The moisture contents at which the equilibrium vapor densities (and hence the volatilization rates) of the residues are reduced to very low values are easily reached in surface soils, drying into moderately dry air. This dependence of volatilization rate upon atmospheric relative humidity and soil moisture content has been observed for trifluralin in the laboratory by Spencer and Cliath,[19] and in the field by Harper et al.,[25] and by Glotfelty et al.[26] Since surface soil temperatures in the field rise rapidly when the surface dries to the point at which heat can no longer be removed

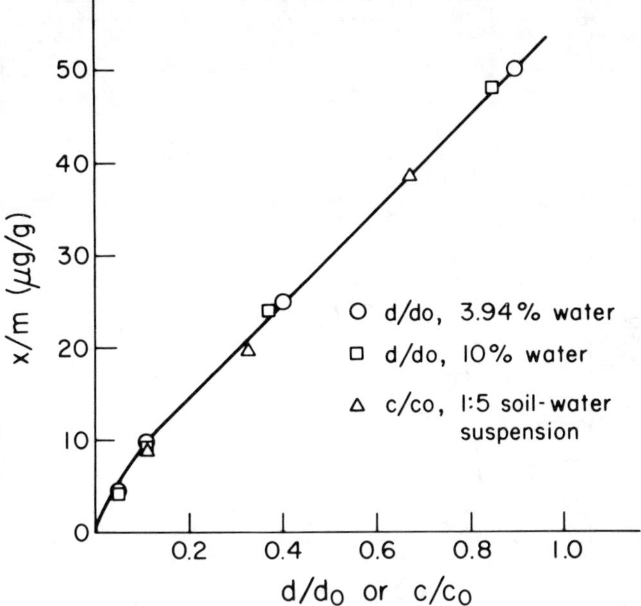

FIGURE 4.   Relative vapor density, (d/d_o), of lindane residues in Gila silt loam soil at 3.94% moisture content and relative solution concentration, (c/c_o), in a 1:5 soil:water suspension as functions of residue concentrations in soil. (From Spencer, W. F. and Claith, M. M., *Soil Sci. Soc. Am. Proc.*, 34, 574, 1970. With permission.)

by the latent heat of evaporation of water from dry soil, where pesticide adsorption will also rapidly increase, no simple correlation between volatilization rates and soil temperature can be anticipated. In experiments using a "microagrosystem" growth chamber to measure the disappearance of trifluralin and several organochlorine insecticides from the surface of bare soil, Nash[27] found that soil moisture was the most important variable affecting volatilization rate, followed by air temperature. The effect of rewetting the soil on the ninth day of this experiment caused a much more dramatic renewal in the volatilization of the insecticides endrin, dieldrin, and DDT than in that of the trifluralin; but this was due to the greater depletion of the trifluralin residues during the first seven days of the experiment. Heptachlor and lindane, which had initial volatilization rates similar to the trifluralin, were also depleted early in the experiment.

This interrelation between temperature, soil moisture content, and the fugacity (or activity) of pesticide residues is thus of major importance in controlling losses of residues from soil surfaces. It is also important in controlling the effectiveness of the herbicides themselves. The field observation that herbicides applied to the surfaces of dry soils are not effective until they have been activated by moisture as rain, heavy dew, or irrigation water is clearly related to their desorption from the strong adsorption sites as the soil moisture is increased.[19]

Early observations of the interaction between the rates of evaporation of water and DDT[28] were interpreted as "codistillation", which implies a bulk flow process taking place when the combined vapor pressures of the two components equal or exceed the ambient pressure. This name is a misnomer and should not be used. Experimental proof that the volatilization of trifluralin and dieldrin from soils is unrelated to the rate of water evaporation has been presented by Spencer[19] and Igue et al.,[29] who both demonstrated that maximum evaporation rates were observed into air at 100% relative humidity, where the net rate of water evaporation was, of course, zero. Pesticides such as DDT, which have fairly high Henry's law coefficients

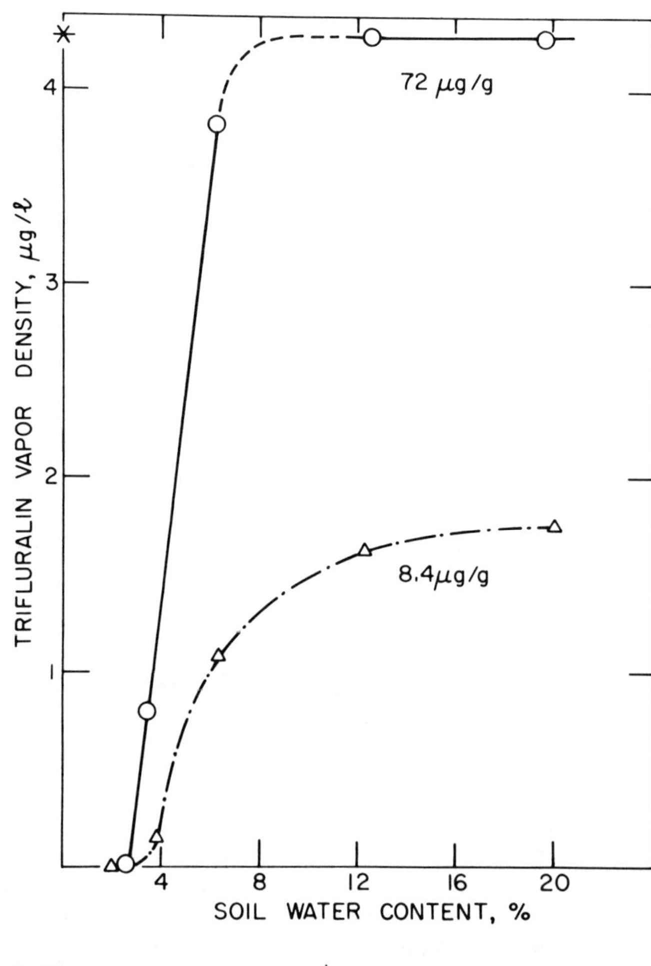

FIGURE 5.   Vapor densities of trifluralin and dieldrin residues in Gila silt loam soil as functions of soil water content. (A) Trifluralin at 8.4 and 72 ppm and 30°C and (B) dieldrin at 100 ppm and 30 and 40°C. (From (A) Spencer, W. F. and Claith, M. M., *J. Agric. Food Chem.*, 22, 987, 1974; and (B) Spencer, W. F., Claith, M. M., and Farmer, W. J., *Soil Sci. Soc. Am. Proc.*, 33, 509, 1969. Both with permission.)

due to their extremely low solubilities, are relatively volatile from water solutions and very small amounts of water will evaporate during the interval during which most of the DDT will volatilize.

5. Soil Organic Matter

The influence of soil organic matter content on the vapor pressure of dieldrin residues in five soils with varied clay and organic matter contents was investigated by Spencer.[14] The results, summarized in Table 3, showed that the vapor densities over both wet and dry soils were inversely related to the soil organic matter contents. The clay content of the soils, which was inversely related to the organic matter content, had only a minor effect. The vapor densities were greatly decreased in the dry soils, but the same inverse relationship to soil organic matter was apparent. These results suggest that clay surfaces do not play a major role in controlling the fugacity of residues of nonpolar compounds under moist con-

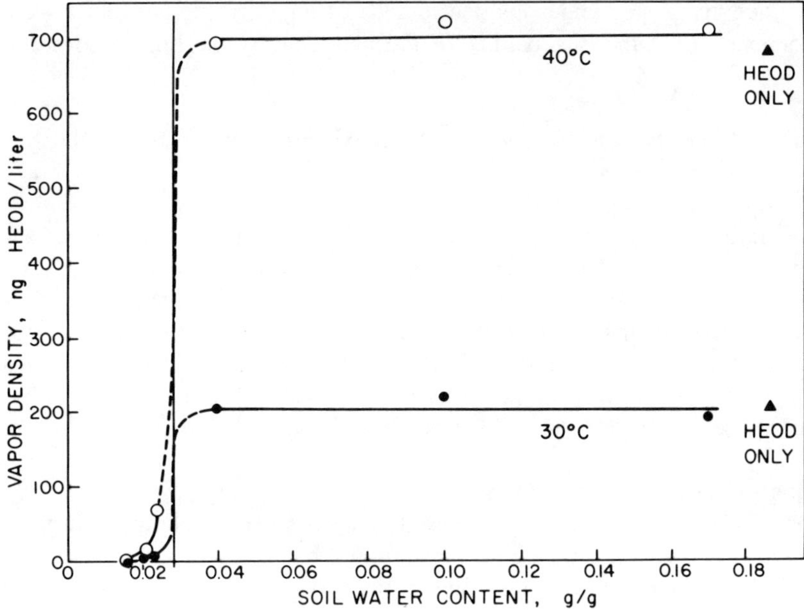

FIGURE 5B.

Table 3
**EFFECT OF ORGANIC MATTER AND CLAY
CONTENT ON VAPOR DENSITY OF DIELDRIN
AT 30°C IN WET AND DRY SOILS
CONTAINING 10 PPM DIELDRIN**

| | Content% | | | |
| | Organic | | Vapor Density | (ng/ℓ) |
Soil type	matter	Clay	Wet[a]	Dry[b]
Rosita vf. sa. l.	0.19	16.3	175	1.7
Imperial clay	0.20	67.6	200	0.9
Gila si. loam	0.58	18.4	52	0.7
Kentwood sa. l.	1.62	10.0	32	0.4
Linne clay loam	2.41	33.4	32	0.6

[a]   Wet - approximately 2 atm. matrix suction.
[b]   Dry - equilibrated at 50% relative humidity.

From Spencer, W. F., *Pesticides in the Soil: Ecology, Degradation and Movement,* Michigan State University Press, East Lansing, 1970. With permission.

ditions, when sorption on or into the organic phase may be the controlling reaction. It has been shown that the variability in the distribution coefficients relating pesticide concentrations to the amounts adsorbed by different soils (defined as $K = c_s/c_1$, where c_s and c_1 are the concentrations in the solid and solution, respectively,) is much less when written in terms of the organic carbon distribution coefficient, $K_{oc} = K/f_{oc}$, where f_{oc} is the fraction of organic carbon in the soil. On this basis, the effects that differences in organic matter, such as those in Table 3, have on volatilization rates can be predicted for many modeling purposes.[30] It is, however, also clear from Table 3 that the effects of changes in soil water content on volatilization rates will be greater than differences in organic matter content. The reduction in biological activity of herbicides applied to soils of high organic content is well known. Both phenomena reflect the inactivation of the herbicide by the high adsorption in such soils.

III. DISTRIBUTION AND PHYSICAL STATE OF RESIDUES

The previous section considered the principal chemical factors that control the vapor pressures of herbicide residues immediately above soil and plant surfaces. In addition to these, a number of physical factors are also of major importance since they control the rate at which the supply of vapor is maintained. The distribution of the residues within the soil is the most important of these because this defines the length of the pathway over which the residues must move by diffusion or bulk flow to the evaporating surface. This distribution is in turn controlled by the way the herbicides are applied and, in some degree, by the type of formulation used. The importance of these factors will be considered in this section.

A. Volatilization of Surface Residues

Where herbicides are applied as water-based sprays, either as solutions, dispersed emulsions, or wettable powders, the residues are initially deposited as a thin layer of material on the exposed plant and soil surfaces. Where this deposit is deep enough that herbicide is lost from the deposit surface while its area does not change significantly, the volatilization rate will be independent of the amount remaining, since it will reflect the "specific volatilization rate" measured in grams per square centimeter of deposit area. The rate will not, however, be identical for different compounds, even if applied together in the same spray, because as indicated by equation 1, the rate of volatilization of each compound will depend upon the product of the molecular diffusion coefficient and the saturation vapor density. Since, as noted by Hartley and Graham-Bryce,[3] diffusion coefficients for gases are approximately inversely proportional to the square root of the molecular weights; differences in volatilization rates from such deposits will reflect this relation.

Eventually the droplet areas will contract in size and, although the specific volatilization rate of the residues may remain constant, the overall rate measured in terms of weight loss per unit of soil or plant leaf area will become proportional to the amount remaining. This indicates that the volatilization rate will decay exponentially with time.[31]

A more complex situation arises when spray formulations are applied to dry, porous surfaces where the droplets are adsorbed into the underlying soil or plant tissue by capillary action. The residues will then remain in a thin (but not necessarily continuous) layer below the exposed surface. It may then be expected that as the residues in the exposed surface are depleted, the volatilization will become dependent upon the rate at which the residues diffuse upward to the free surface as a concentration gradient is established. The volatilization rate will then be controlled by the "back-diffusion" of the residue. The way that the volatilization rates that are controlled by diffusive flow change with time have been analyzed by Mayer et al.,[32] who solved the flow equations for five model systems, each with differing boundary conditions. These analyses indicate that in moist soil where there is no bulk transport of residues in moving soil water, the volatilization rates should decrease approximately in proportion to the square root of time, with the exact relationship depending upon the boundary conditions. This prediction was verified in laboratory measurements of volatilization rates of dieldrin[32] and lindane[33] residues mixed into the top 5-mm depth of moist soil; a similar laboratory test of the volatilization of incorporated triallate was reported by Jury et al.[34]

Validations of the predictions of such models in the field are very difficult, owing to the complex effects of residue distribution, sampling errors and the uncertainties about boundary conditions; but the type of relationship predicted was observed by Glotfelty et al.[26] in an experiment on the volatilization of surface applied trifluralin, heptachlor, chlordane, and lindane from moist soils in the field.

Since the increased adsorption in dry soils reduces the vapor pressures to such low values that volatilization effectively ceases whenever the topmost layer of the soil surface becomes dry, the predicted relationships can only be expected to be found over time periods where the soil remains moist. This reduction in fugacity by increased adsorption in dry soils can become the dominant factor controlling the volatilization loss. This effect is also of major importance in that it negates any simple relationship between volatilization rate due to changes in temperature because marked increases in soil surface temperature only occur after water evaporation has ceased and the soil surface is dry. Both aspects of increased adsorption are strikingly demonstrated in the data obtained by Turner et al.[35] on the volatilization of surface-applied chlorpropham.

It is thus evident that in addition to differences in chemical properties, the volatilization rates of surface-applied herbicides in the field may also be expected to reflect the physical distribution of the residues on and within the target surface in a very complex manner depending upon residue penetration and coverage and the moisture status of the surface. In field situations, the expression of these physical effects will reflect herbicide application and management techniques and, except where different chemicals are compared under identical usage patterns, may obscure differences in volatilization rates due to differences in their chemical properties.

B. Volatilization of Incorporated Residues

Where herbicide residues are not left exposed on soil and plant surfaces, but are incorporated into the soil by cultivation, the rate at which they volatilize becomes controlled by their movement through the body of the soil to the overlying air. Incorporated residues differ from surface applications in two respects. First, as noted in Section II., the average concentrations in the soil are lower by about one order of magnitude because they are mixed into the soil to depth of 5 to 8 cm (about 2 to 3 in.) as opposed to the 3- to 5-mm depth to which formulation spray drops are likely to penetrate. As a result of this lower concentration, the equilibrium vapor pressure of these residues will be considerably lower, as illustrated in Figure 2B for trifluralin and discussed in Section II.B.1. Second, the average diffusion path length to the surface will be longer by at least the same factor, so that in addition to the decrease in fugacity, the resistance to flow will be much greater. Thus, except for the small fraction of exposed material remaining on the surface that is readily lost, the volatilization of incorporated residues will be greatly restricted and be dependent upon the rate of their upward movement to the soil surface.

The factors controlling this movement have been discussed in detail by Spencer et al.,[36] who contrasted the volatilization patterns to be expected when the flow to the surface is controlled by simple diffusion gradient with those expected where pesticide is transported to the surface in the upward flow of soil water (convective flow).

1. Diffusion Controlled Flow

After the initial uniform incorporation into a moist soil, the herbicide concentration in the soil solution is controlled by desorption from the solid phase, as illustrated in Figure 6A. Since the vapor density in the overlying air (or within the air-filled soil pores) is in turn controlled by the Henry's law distribution, this vapor density will reflect the adsorption-desorption reactions. This can be interpreted as the effect of the adsorption in reducing the fugacity of the residues. In the absence of upward evaporative movement to the surface, as when the relative humidity of the air is close to 100%, the herbicide will volatilize from the surface until the diffusive flow becomes limiting (Figure 6B); this situation is described mathematically by model II in the analyses presented by Mayer et al.[32] This model has been tested and verified in laboratory measurements of lindane and dieldrin volatilization reported by Igue et al.[29] and by Spencer et al.[36,37] These studies, and others by Jury et al. for triallate[34]

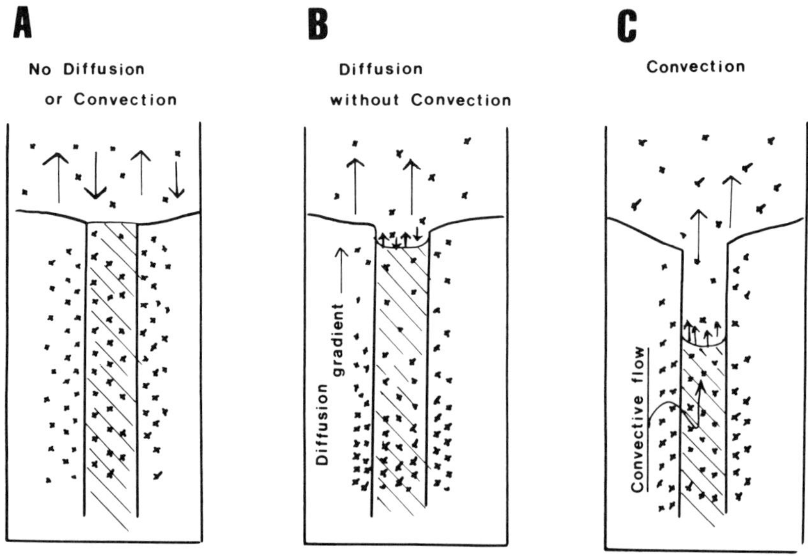

FIGURE 6.   Diffusive and convective (wick effect) flow mechanisms controlling volatilization rate of herbicides incorporated into soil.

and by Spencer and Cliath for trifluralin,[19] suggest that time periods of up to one month are needed for steady state diffusion gradients to be established. Both the time required for this and the final rates of diffusive flow will reflect the overall diffusion coefficient of the pesticide in the particular soil. Since this coefficient is a complex function of the diffusion pathways through the solid, liquid, and vapor phases in the soil, and these will change with the water content, no single value of such an "effective diffusion coefficient" can be expected to characterize the behavior of even a single pesticide in one soil. Predictions of diffusive flow must rely upon values obtained by calibration observations with the chosen combination of pesticide and soil. Such "effective diffusion coefficients" may be expected to be sensitive to soil bulk density and temperature in addition to soil water content. Where, however, the soil moisture content is reduced below the critical values discussed in Section II.2., and the equilibrium vapor densities are reduced to very low values by the adsorption in dry soil, diffusive movement will be almost wholly inhibited. In terms of the fugacity approach, this means that this function is also reduced to a very low value and any layer of dry soil that may form on the surface represents a section of the dispersal pathway with a prohibitively high impedance to flow. These conditions can of course be frequently found in the field, where fallow surface soils in the field - and particularly those of light texture - are exposed to intense drying by wind and sun.

2. Convective Flow

Where continuous upward movement of water is induced by evaporation, the resulting upward transport of pesticide may prevent or distort the development of a diffusion gradient, and volatilization at the surface will continue with the water movement (Figure 6C). This flow mechanism has been described by Hartley[38,39] as the "wick effect" by analogy with capillary liquid flow to the ends of exposed fibers. Since, as demonstrated by Spencer and Cliath,[37] the pesticide and water evaporation rates are independent, this is not a true "co-distillation"; the rate limiting step is the supply of pesticide to the surface from the reserve of residue adsorbed on the solid surfaces within the soil pores.

The relative importance of diffusive and convective flows are difficult to assess because diffusion is always present during the convection process. Laboratory data for lindane and dieldrin show that with these insecticides volatilization supported by convective flow in moist soil can be up to five times faster than that controlled by diffusion. Similar experiments with trifluralin[19] showed much smaller differences reflecting the slower transport of trifluralin to the surface due to its lower concentration in the soil water. Similar results were obtained in measurements of triallate volatilization by Jury et al.[34] who showed that the relative contribution from convective flow was much less in a silt loam than a sandy loam soil, owing to the lower solution concentration established by the greater adsorption in the silt loam, which also had a much higher organic matter content.

Upward movement of moisture to the soil surface is not, however, a continuous process under field conditions, and may cease during rainfall or become very slow at high relative humidity and when there is no insolation to supply the latent heat of evaporation of the water. It may therefore vary greatly during the day, and will almost always be very small at night. The convective flow of pesticide to the surface will show similar variations, and while it may be very difficult to use the data obtained under highly controlled laboratory conditions to predict the quantitative effects of convective flow in the field, the concept has proved essential for the interpretation of a number of the field observations discussed in Section V.A. It should, however, be noted that, in common with diffusive flow, convection controlled volatilization of pesticide will almost totally cease when the rate of upward movement of soil water is less than the rate of water evaporation so that the surface soil becomes extremely dry. As long as the lower layers of soil remain moist, water evaporation will continue, and both water and pesticide will continue to be moved to the surface where the pesticide will be adsorbed and accumulate in the dry soil layer. Rewetting of this layer by rainfall, irrigation, dew or increased humidity will then permit increased volatilization until the surface accumulation is depleted. This effect was demonstrated in the laboratory with dieldrin residues by Spencer and Cliath.[37]

The pattern of water loss from soil is, therefore, an important variable that must be taken into account in interpreting observed volatilization losses in the field or in developing mathematical models to describe or predict them. Calculations by Spencer and Cliath[37] showed that volatilization rates of lindane and dieldrin residues could be interpreted in terms of measured water loss and the concentration of the insecticides in the soil solution. A model describing the effects of adsorption, air-water distribution, soil bulk density, and water flow was presented by Jury et al.[34] and validated by laboratory measurements. A more comprehensive model, describing transport and loss of soil-applied pesticides (and other organic chemicals) by various pathways, and designed specifically to predict the comparative behavior of different materials, has been developed by Jury et al.[8] The significance of these and other interpretations of the mechanisms controlling volatilization losses will be examined below since they can only be finally evaluated by comparison with field measurements of volatilization or herbicide disappearance.

IV. METHODOLOGY FOR VOLATILIZATION RATE MEASUREMENTS

Experimental systems have been developed for measurements of volatilization rates from soils on both laboratory and field scales. Laboratory systems may be divided into experiments using small amounts of soil in confined and highly controlled conditions as described by Farmer et al.[33] and by Spencer and Cliath,[37] and the growth-chamber scale "micro-ecosystems" described by Nash and Beall.[27,40] The type of data obtained from both classes of laboratory systems differs from those obtained in the field measurements because the latter require observations of a different character taken on a much larger scale. Laboratory experiments can measure "steady state" volatilization rates established when temperature, humidity, soil moisture, and air flow speeds are maintained constant over extended periods of time or, alternatively, they can measure the changes in rate of loss as the residues are depleted under constant conditions. The greatest value of such laboratory studies lies in the ease with which they can be used to compare the behavior of different chemicals or formulations under similar environmental conditions and to provide data to assess the probable importance of these differences in the field.[27,36]

Field measurements of volatilization rates are usually based upon upward flux rates calculated from gradients of vapor concentration in turbulent air flowing over treated fields measured over periods of 30 min or longer. The results obtained reflect the natural, uncontrolled soil and weather conditions which can of course themselves show rapid changes. Since the sampling periods are relatively short, such data can sometimes reveal very clearly the effects of such changes on the volatilization rates. The principal disadvantages of field experiments lie in the elaborate instrumentation required and the large number of samples that must be analyzed. These, together with the theoretical approach necessary for calculation of flux rates, will be discussed below.

A. Disappearance Methods

Where the experimental conditions are such that the disappearance of a herbicide applied to a soil can only be attributed to either chemical degradation or volatilization, an estimate of the latter can be made by measurements of the rate of disappearance if the rate of chemical degradation is known. The apparent advantage of this approach is that it avoids vapor sampling and analysis and does not require elaborate experimental control systems. If the measurements are made over a short enough time so that it can be assumed that both processes can be described by "first order" chemical kinetics, and there are no major changes in the temperature and moisture regimes so that both rates can be taken as uniform during the experiment, the disappearance of the herbicide may be approximated by the kinetic equation describing simultaneous reactions, $R_t = R_o \cdot - \exp(k_c + k_v)t$, where R_t and R_o are the residue concentrations initially and at time t, and k_c and k_v are the coefficients for the rates of chemical degradation and volatilization. Subject to these assumptions, the value of k_v may be obtained by difference from the measured rate of disappearance and the value of k_c where this is known from measurements in closed systems where volatilization is experimentally prevented.

In practice, the method can only be expected to give consistent results where the rate of disappearance is measured while the evaporation is not affected by changing temperature and moisture conditions, and the residues are not seriously depleted. These restrictions imply that the technique is only likely to be successful with small samples under controlled conditions in the laboratory or in ecosystem chambers with elaborate control systems, or with highly volatile compounds so that measurements can be made over short time periods. When used in the field, the method is subject to great uncertainty due to sampling variability which may impose limits of up to $\pm 30\%$ on individual estimates of R_t.[41,42] Except where the volatilization loss is much larger than the degradation rate, the value of k_v may be less than the error associated with the measured value of the overall disappearance coefficient $k_t = (k_c + k_v)$. This again implies that the method is only likely to prove satisfactory in the field for pesticides of relatively high vapor pressure, and where moisture and temperature conditions will support rapid volatilization loss. Although the method appears simple in principle, in practice it involves a number of uncertainties that make it unsuitable except as a means for making approximate comparisons between the volatilization of different compounds under strictly comparable conditions.

B. Vapor Flux Measurements

Measurements of pesticide volatilization in the field are made by calculation of the flux of pesticide vapor through a horizontal plane in the turbulent boundary layer of the atmosphere overlying the crop canopy or treated soil. This requires the measurement of the gradient of pesticide concentration with height in the air over the field. This is done by sampling the air at a series of heights (usually up to about 2 m) above the soil surface for chosen time periods; these are commonly between 30 min and 2 or 3 hr, depending upon the expected weather conditions and the pesticide concentrations. Measurements of gradients of temperature, water vapor concentration, windspeed, and other supporting meteorological variables must also be made over the same time periods. Care must be taken to ensure that the mast is placed so that it is entirely within the atmospheric boundary layer characteristic of the area treated with pesticide. Since the slope of the upper edge of this layer is often as low as 1%, this means that a mast of 2m in height must have a clear fetch of 200 m to the upwind edge of the treated zone to ensure that the characteristic profiles are obtained.

1. Theoretical Principles

The principles used in the calculation of the pesticide flux have been presented by Parmele et al.[43] This calculation is based upon Equation 2. Since there are no sources or sinks for the vapor in the air above the soil or crop canopy, the upward flux through the turbulent air will be uniform with depth and independent of height. (It should be noted that this assumption may not be valid in air under a crop canopy, where there may be release or adsorption of pesticide by leaves. A more complex approach than that presented here must therefore be used for pesticide flux between the soil surface and the crop canopy.) The value of dp/dz at the chosen height z is experimentally accessible from the vapor concentrations measured over a range of heights, but the value of K_p is not known and must be inferred from other measurements.

To this end it must be noted that Equation 2 is one of a general set of equations of the form $F = k_z \cdot (dx/dz)$ used in micrometeorological studies to describe the vertical transport of water vapor, heat, momentum, and carbon dioxide or other gases through the turbulent boundary layer. Since this transport is outside the laminar boundary layer, molecular diffusion effects are discounted and the dilution of each component is controlled by the mixing in the turbulent air flow. As a first approximation, the values of k_z in each of the appropriate forms of the equation may then be assumed to be the same (the similarity principle). If independent measurements of the flux of a particular component ($\uparrow$ F) and the corresponding gradient (dx/dz) can be made, the value of K_p may be calculated and used in Equation 2 to calculate the pesticide flux ($\uparrow$ P).

The assumption that K_p may be identified with k_z values for water vapor, heat flow, and momentum exchange is not equally good in all cases. Although detailed examination of this topic is beyond the scope of this review, some comment is necessary as a background to the commentary on the field data presented in Section V.A. Excellent detailed discussions have been published by Lemon,[44] Parmele et al.,[43] and Montieth.[45]

Of the three approaches - water vapor flux, energy balance (or heat transport), and aerodynamic (or momentum flux) - the first is probably subject to the fewest assumptions. Since the upward flow of water evaporating from the soil or plant surface is a molecular process parallel to the volatilization and dispersal of the pesticide molecules, both may be expected to be diluted in the same way by the turbulent mixing of the air. The principle difference between the two flow patterns is likely to be due to the influence of the background concentration of water vapor in the atmosphere which will cause a different gradient in the upper part of the profile as it asymptotically approaches the background level; no similar background exists for the pesticide. Any ambiguities resulting from this difference can be minimized by confining flux and gradient measurements to the lower regions of the boundary layer. The gradient of water vapor concentration is calculated from the humidity gradient measured at chosen heights. The need for companion measurement of water vapor flux imposes a considerable restriction on the possible use of this method because it requires direct measurements of evapotranspiration from the soil or crop surface. Since this can show rapid fluctuations in response to changes in insolation, it must be measured as accurately as possible over periods coincident with the pesticide sampling times. This essentially restricts the use of the method to locations equipped with recording lysimeters. Few measurements of pesticide volatilization using this approach are reported in the literature; the most notable example is the data obtained by Parmele et al.[43] and by Taylor et al.[46] for dieldrin and heptachlor volatilization at a hydrologic research station in Ohio.

The energy balance, or Bowen ratio method, described by Parmele et al.,[43] Lemon[44] and Rose,[47] was originally developed to calculate the water vapor flux in the absence of lysimeter data. The flux of water vapor, E, is calculated by partitioning the dispersal of energy from the soil surface according to the following equation:

$$R_n = G + H + (L \cdot E) \tag{5}$$

where R_n is the net incident radiation, G the soil heat flux, H the upward transport of sensible heat and L·E the heat carried by the dispersing water vapor, L being the latent heat of evaporation. On introducing the Bowen ratio, H/(L·E), Equation 5 transforms to

$$\uparrow E = (R_n - G)/L \cdot (1 + B) \tag{6}$$

where $B = y \cdot \Delta T/\Delta q$, ΔT and Δq being the differences of mean air temperature and water vapor density over a selected height interval within the profile; y is the psychrometric constant. Using the value of $\uparrow E$ obtained in this way, the value of K_e may be obtained from the water vapor flux equation $K_e = \uparrow E \cdot (dq/dz)$ and, together with the measured pesticide vapor concentration gradient dp/dz, used to calculate the pesticide flux from Equation 2. It should be noted that the similarity principle assumption enters this calculation at two points, the value of k_z being assumed to be the same for both pesticide and water vapor and sensible heat flux; the latter assumption is implicit in the use of the ratio $\Delta T/\Delta q$. The principle disadvantage of the energy balance method is in the complexity of the supporting data. This must include measurements of net radiation, soil heat flux, and temperature and humidity profiles, requiring a considerable amount of equipment with significant maintenance and calibration costs, together with mobile power supplies and shelter in the field. Although a number of sets of field data on pesticide volatilization based upon this method have been published, the need for complex equipment has undoubtedly precluded its wider use.

The aerodynamic method requires measurements of the gradient of wind speed with height above the crop. Wind speed increases with height to the value imposed by the broad scale meteorological pressure patterns. The decrease in velocity in the boundary layer close to the ground reflects the friction or drag exerted by the soil or crop surface as the momentum in the flowing air is transferred to the surface. This downward flux or diffusion of momentum can be described by an equation similar to those for heat, water vapor, and other gases:

$$Fm = -K_m \cdot (du/dz) \tag{7}$$

where K_m is the eddy diffusivity coefficient for momentum transfer at the height z, and du/dz is the gradient of windspeed with height. In principle, it is clear that if we equate the diffusivity coefficients K_p and K_m, the pesticide flux P can be derived from measurements of pesticide vapor concentration and wind speed over suitable ranges of height. The conditions under which this equality may be assumed have been discussed by Montieth[45] and Rose.[47] The assumption implies that the exchange of momentum, heat, and vapor are all due to turbulent mixing of the air by mechanical forces of surface friction and drag, and that the upward or downward transfers of mass or momentum caused by buoyancy due to heating can be neglected. In turn, this means that the equality can only be expected to hold when the atmosphere is in neutral stability with an adiabatic lapse rate; the profiles of water vapor, pesticide concentration, and momentum above the surface should then all have similar geometrical shapes. Departures from neutral stability are imposed when the rate of surface heating is greater than the upward flow of turbulent convection so the "bubbles" of warmer air move upward transporting mass and energy and distorting the wind-induced turbulent flow, or under radiative cooling when the surface layers become dense and resist the stirring action of the wind. These effects of buoyancy are corrected for in part by the use of the Richardson number R_i, which is defined in terms of the ratio of the temperature lapse rate and the wind velocity gradient.[47] The final form of the equation used in the calculation of the pesticide flux using the aerodynamic method reduces to

$$P = k^2 \cdot \frac{(C_1 - C_2)(U_2 - U_1)}{\theta^2 \cdot [\ln(Z_2 - Z_0)/(Z_1 - Z_0)]^2} \tag{8}$$

where C_1, U_1 and C_2, U_2 are the vapor concentrations and wind speeds at the heights Z_1 and Z_2 above the surface, and k is the Von Karman constant, assumed to be 0.4. Following Parmele et al.,[43] the stability correction term may be defined as $\theta_2 = (1 - b \cdot R_i)^{-0.25}$, where R_i is the Richardson number and b is a constant (close to 18) dependent upon the roughness of the surface. The aerodynamic method therefore requires the measurement of the pesticide concentration, wind profile, and temperature gradients over a common range of heights. The relative simplicity of this experimental requirement has made this the method of choice in a large number of published field experiments, despite the possible ambiguities in the assumptions upon which it is based. The extent to which these throw doubt upon the measured volatilization rates must, however, be judged by inspection of the individual experiments. The assumptions are probably least trustworthy where measurements have been made under vigorous thermal heating or very stable conditions in the morning and evening.

The choice of the method used in a particular experiment is often dictated by the resources available. The use of the direct water vapor flux method is highly restricted because it requires an accurate and sensitive lysimeter to measure evapotranspiration from the same crop or soil surface to which the pesticide has been applied. The energy balance method, in which the water vapor flux is calculated from the Bowen ratio, has the theoretical advantage that there probably are less uncertainties in equating the values of K_e and K_p than with the momentum approach, but the equipment needs are much more elaborate.

The energy balance method is perhaps to be preferred where fluxes and evapotranspiration rates are likely to be high. The method, however, has the disadvantage that the calculated values of K_e may be indeterminate in the morning and evening or under cool conditions when the net energy input ($R_n - G$) is small. The method can also give difficulties where the energy balance requires terms for advected energy as well as radiation and conduction; this most typically happens in the "oasis situation", where warm, dry air blowing into the experimental area causes water evaporation in excess of that measured as R_n and G. These conditions were encountered by Cliath et al.[48] in measurements of EPTC volatilization from irrigated fields in southern California, when the water evaporation caused stable atmospheric inversions over the plot at midday.

In contrast, the aerodynamic method requires less equipment, but is theoretically less reliable when wind speeds are low and thermal turbulence is dominant; its value in the morning and evening (when pesticide volatilization rates can sometimes be large) depends upon the windspeed gradient. When this is maintained outside daylight hours, the aerodynamic method may give valuable results when other methods fail.

A few observations of pesticide fluxes that permit direct comparisons between the various methods have been made,[25,26,49] but the fundamental question of the relationships between the values of K_w, K_e, and K_m has been discussed at length in the micrometeorological

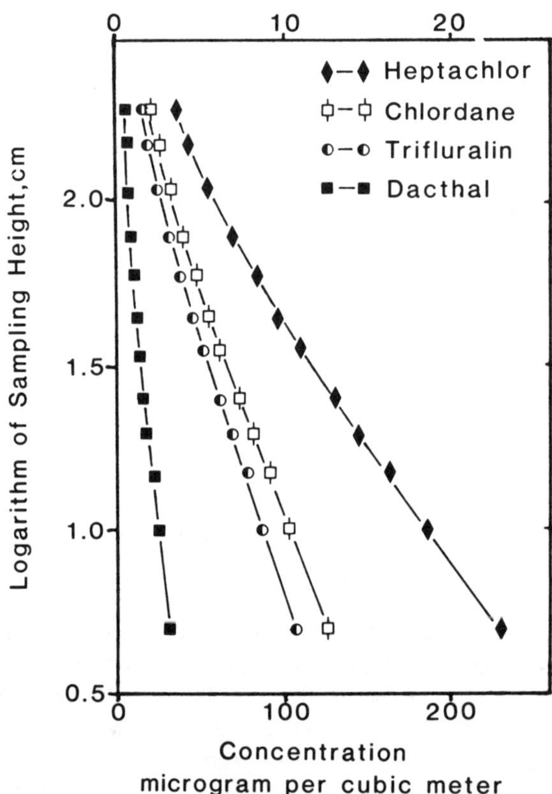

FIGURE 7.   Vapor concentration profiles of heptachlor, chlordane, trifluralin, and dacthal up to 120 cm height after surface application to bare, moist, silt loam soil at Beltsville, Md., in August 1975. (From Glotfelty, D. E., Taylor, A. W., Turner, B. C., and Zoller, W. H., *J. Agric. Food Chem.*, 32, 638, 1984. With permission.)

literature.[44,45,47] In the few cases where comparisons are possible, the data indicate that the differences between them tend to vary with atmospheric stability and the character of the soil or crop surface.

Relative rates of volatilization of two herbicides can be determined from the slopes of their profiles (dp/dz), provided that these are measured in the same samplers over a plot to which both herbicides have been applied together as in the example presented in Figure 7, which compares simultaneous profiles of heptachlor, trifluralin, chlordane, and dacthal.[26,50] Such simultaneous profiles measured at different times may also give much insight into the response of different compounds to changing soil moisture, surface temperature, and wind conditions. The main difficulty caused by the uncertainties in the absolute rates - which cannot be calculated without the meteorological data - will arise when field volatilization rates are used to estimate the contribution of the volatilization to the disappearance of the residues.

It should also be clearly recognized that simply measuring pesticide concentrations in air, or even profile gradients of a single pesticide at different times without meteorological data, even at the same location, gives no information about volatilization rate because of unknown changes in K_z; such changes in K_z can sometimes extend over several orders of magnitude.[47]

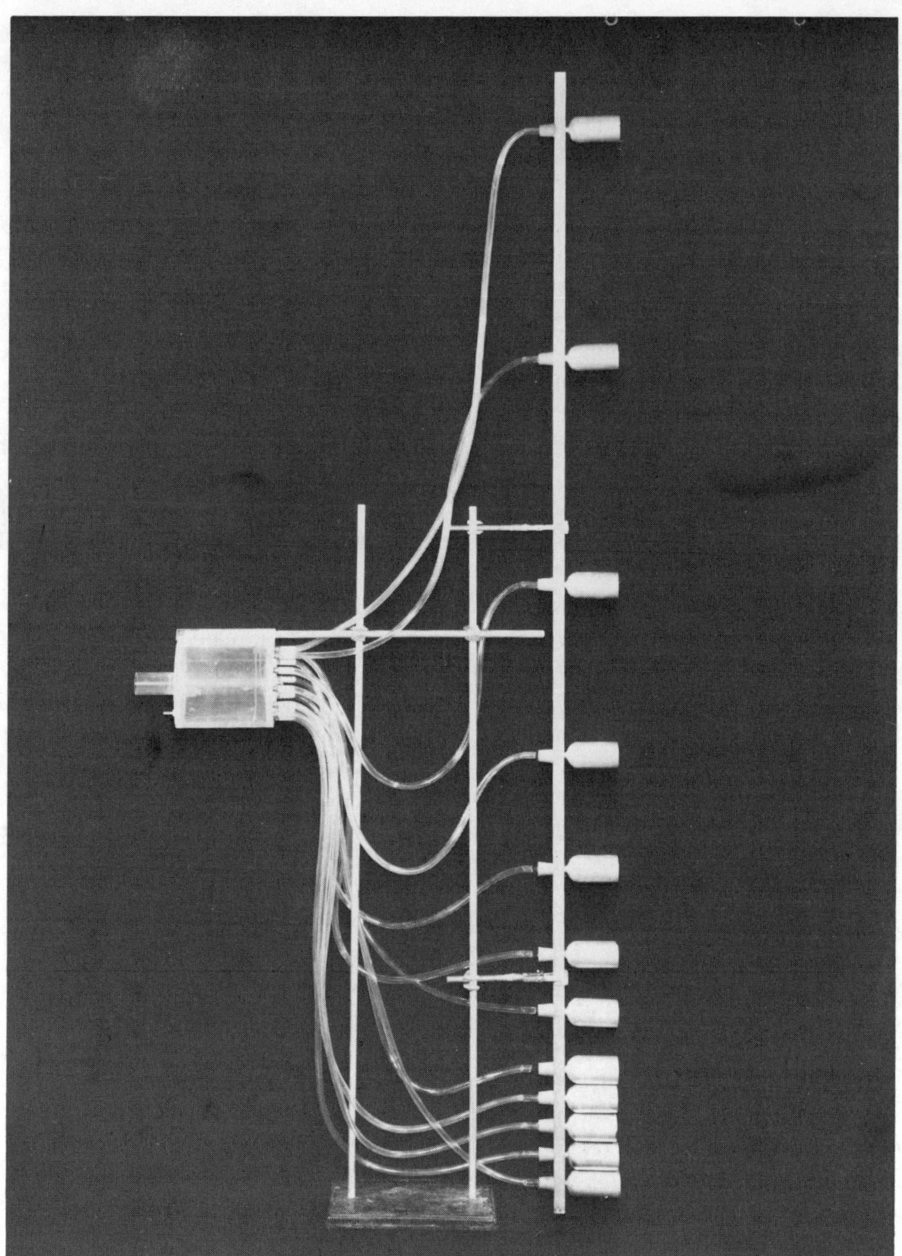

FIGURE 8.   Air sampling mast for measurement of herbicide vapor profiles over treated surfaces. Mast carries twelve sample ports feeding into single presure surge control tank, left.

2. Air Sampling Techniques

The profile of pesticide vapor pressure over a treated area is measured by taking samples of air with samplers mounted on vertical masts at a chosen series of heights, as illustrated in Figure 8. Each sampler consists of a glass or metal tube containing adsorbent which removes the pesticide vapor from the air drawn through it at a predetermined rate. In the equipment shown in Figure 8, the individual samplers are attached to a single surge tank to control and balance the overall airflow through the whole set of samplers, the surge tank being evacuated by a high-volume suction pump. Commercial shop-type vacuum cleaner pumps have been found very suitable for this purpose, producing stable airflow rates over periods of several hours. The rate of air intake through each sampler depends to some extent upon the resistance of the adsorbing material, and must be measured with a flowmeter at several times during each run. This measurement is essential to derive the concentration in the air from the amount trapped by the sampler.

Since the decrease in pesticide concentration with height is logarithmic, it is convenient to space the sample heights exponentially, as indicated in Figure 8, to simplify data interpolation. Sampling times can be varied depending upon the experimental objectives, but a minimum of 30 min appears to be desirable to avoid irregularities due to the random turbulent vapor flow. While periods of more than 2 hr may sometimes be necessary to collect enough vapor for accurate analysis, longer collection times are undesirable because other factors, such as windspeed, temperature, humidity, and soil moisture, may be changing so quickly that steady state conditions cannot be assumed over the observation period.

A range of materials may be used as the adsorbing substrate in the samplers. In many early experiments, air was pulled through bubblers containing hexylene or ethylene glycol. This method was however discarded in favor of solid adsorbents because of the difficulties of transferring liquids in the field, and because the adsorption of water by the glycol caused changes in liquid volume and flow rate in all but the driest atmospheres. Solid adsorbers, such as polyurethane foam, offer great convenience in transfer and storage of samplers, but require elaborate cleaning and purification before use.[51] For many purposes, powdered materials, such as Chromasorb® or XAD® resin, may be used; these require less chemical preparation, but, as powders, they require more careful handling in transfer and filling operations and in controlling the density of packing in the sampler, which may affect resistance to air flow. Important considerations in the choice of material include the stability of the residues on the adsorbent during sampling and storage, and the ease with which the residues may be recovered during the analysis. The range of materials available for use as adsorbents has been reviewed by Bidleman.[52]

V. FIELD MEASUREMENTS

It is clear from the earlier discussion that a number of interactions between chemical properties and soil factors are likely to be important in controlling the evaporation of herbicides from treated areas in the field. Predictions of their relative importance may, however, require simplifying assumptions and major extrapolations about the complex and often transient conditions of the uncontrolled field environment. Field measurements of volatilization rates, supported by data on herbicide usage and management, residue concentrations, soil moisture levels, surface temperatures, and weather conditions are therefore an essential element in understanding both the rates of dispersal of the residues into the environment, and the way in which these are affected by their chemical properties and soil factors. Without such field data, no real calibration of laboratory data or modeling projections can be made.

The number of such experiments is, and probably will remain, rather limited. They are complex and expensive, requiring the collaboration of meteorologists, pesticide chemists, weed scientists, and field staff; equipment needs are also demanding, requiring sensitive meteorological instruments and laboratory facilities for the rapid and accurate analyses of a large number of soil and air samples. Experiments must therefore be designed to yield the maximum amount of information and provide data to evaluate the important factors in the particular situation. Flexibility in experimental plans is important because the time span during which the herbicide can be applied is often limited by the crop management practice, and once the application is made, the experimenter has no further control over the course of events. Since chemical analyses are never available immediately, decisions about changes in sampling patterns or protocol must be made by judgment without data, and be aimed to obtain the best information possible under the circumstances.

Despite these difficulties, a significant background of field data has been obtained in recent years which confirms and illustrates the importance of the factors identified in the laboratory work. A catalog of the available data from such field experiments with herbicides is presented in Table 4. For comparative purposes, this also includes experiments with the insecticides heptachlor and lindane, whose chemical properties are comparable to those of some herbicides.

A. Review and Interpretation of Field Data

Many of the original experimental reports contain much useful and valuable data that cannot be summarized here. In order to relate this information to the chemical properties, soil conditions, and management practices discussed above, the experiments will be reviewed in terms of several separate topics. These will include (1) long-term volatilization in relation to usage and chemical properties; (2) the effects of soil moisture on volatilization rates; (3) direct comparisons between different compounds; (4) the effects of special utilization patterns; and (5) the identification of the rate-controlling steps in the dispersal process.

1. Long-Term Volatilization

The importance of how a chemical is used is emphasized by the low volatilization rates of trifluralin and heptachlor when these are incorporated to depths of 2.5 cm or more.[46,55] Although, as shown by the data in Table 1, these are among the more volatile compounds, incorporation reduces their loss rates to 10% or less over periods of 90 to 100 days or more. Other data in Table 4 shows the rapid dispersal of both compounds when residues remain exposed on the surface of moist soil.[26]

Since volatilization measurements cannot be made continuously over extended periods of time, long-period estimates are based upon interpolated data for days when observations are not available. In the experiment reported by White et al.,[55] 79% of the trifluralin volatilized was lost in the first 18 days; the weather data presented suggest that the rate showed a steady decline without erratic variations so that the interpolations can be made with some confidence. Since in this experiment the decrease in the residue concentration by chemical degradation was comparable to the volatilization loss, the change in volatilization rate in this experiment can be attributed in large part to disappearance of the residues. A detailed interpretation of the daily data was presented by Harper et al.[25] Data obtained by Taylor et al.[63] on a heavier textured soil under cooler and wetter conditions in New York indicated much smaller and more uniform losses of incorporated trifluralin with a total growing season volatilization of less than 3.4% The total disappearance of residues was about 50% over the same period; this volatility loss was less than the 90% uncertainty level in the value for the overall disappearance over the same time.

The field data for heptachlor reported by Taylor et al.[46] showed a more complex pattern, but since the heptachlor flux was found to be related to the daily evapotranspiration measured in a continuously recording lysimeter available at this site, these could be used to support the interpolations.

In general, the reliability of long period measurements of volatilization depend upon the uniformity of weather and soil conditions prevailing during the experiment. More reliable observations are likely where the greater fraction of the residues are lost during the earlier stages, so that any interpolation errors only affect the estimates of a small fraction of the total residues volatilized; all the available data indicate that this is a common pattern, with early large losses reflecting the dispersal of the part of the residues remaining on or close to the soil surface.

2. Soil Moisture Effects

The influence of soil moisture content is clearest in the data presented in Table 4 for several surface applied materials on dry and moist soils. The tabulated data gives direct

Table 4
FIELD MEASUREMENTS OF HERBICIDE VOLATILIZATION RATES

Compound and useage	Surface	Season and place	Fraction volatilization in period	Ref.
Alachlor				
Surface	Fallow silt loam	May Maryland	26% in 24 days	53
Atrazine				
Surface	Fallow silt loam	May Maryland	2.4% in 24 days	53
Chlorpropham				
Surface	Soil under soybeans	May—July Maryland	49% in 50 days	35
Microencap	Soybeans		20% in 50 days	35
Dacthal				
Surface	Moist fallow silt loam	August Maryland	2% in 34 hr	26
EPTC				
Surface in irrigation water	soil under 25-cm alfalfa	May California	74% in 52 hr	48
Heptachlor				
Incorp. to 7.5 cm	Soil under maize	May—October Ohio	7% in 170 days	46
Surface	Moist fallow silt loam	August Maryland	50% in 6 hr; 90% in 6 days	26
Surface	Dry fallow sandy loam	June Maryland	40% in 50 hr	26
Vegetation	Short orchard grass	July Maryland	90% in 7 days	54
Lindane				
Surface	Moist fallow silt loam	August Maryland	50% in 6 hr; 90% in 6 days	26
Surface	Dry fallow sandy loam	June Maryland	12% in 50 hr	26
Simazine				
Surface	Fallow silt loam	May Maryland	1.1% in 24 days	53
Trifluralin				
Incorp. to 2.5 cm	Soybeans on sandy loam 0.6% o.m.	June—October Georgia	22% in 120 days	55
Incorporated to 7.5 cm	Soybeans on loam 4% o.m.	May—September New York	3.4% in 90 days	63
Surface	Dry fallow sandy loam	June Maryland	25% in 50 hr	26
Surface	Moist fallow slit loam	August Maryland	50% in 7.5 hr; 90% in 7 days	26
2,4-D Octyl ester[a]				
Surface	20-cm-high spring wheat	June Saskatchewan	21% in 5 days	49

[a]   Volatile ester only — does not include acid.

comparisons for trifluralin, heptachlor, and lindane; the most satisfactory comparisons are those measured over periods of a few hours because these are not complicated by the effects of soil surface drying. The detailed data in the original reports, however, provide a number of excellent examples of the effects of changes in soil moisture upon volatilization of both incorporated and surface-applied materials.

a. Incorporated Residues

The effects of soil moisture upon trifluralin volatilization from a Cecil sandy loam in Georgia were described by Harper et al.[25] and White et al.[55] In this experiment, 95% of the trifluralin was contained in the top 2.5-cm layer of soil. Measurements were made of the water content of this layer throughout the experiment, so that volatilization rates could be directly related to the soil moisture levels. Major differences were found in the diurnal volatilization patterns as the soil moisture changed. As the surface soil dried to a water content of less than 0.01 cm^3/cm^3 between 4 a.m. and noon on the second day of the experiment, the volatilization rate fell from 1.7 to below 0.15 g/ha/hr, despite a fourfold increase in the atmospheric diffusivity coefficient, K_z. This soil moisture level corresponds to less than the monomolecular layer level at which the vapor density of trifluralin would be reduced to a very low level by adsorption.[19] As the surface soil moisture content increased in the evening due to the upward movement of sub-surface soil moisture, the volatilization again increased. Very similar results were also obtained on the 18th day, when the soil was again dry and the volatilization fell below 0.04 g/ha/hr during the midday sampling period. On the sixth day, a strikingly different pattern was observed, with a sharp increase in volatilization immediately after the soil was wetted by a rain shower during the midday sampling period. These results clearly indicate the importance of adequate soil moisture for the volatilization of incorporated herbicide residues, as also indicated by the laboratory results obtained by Spencer and Cliath.[19,37] The effect clearly reflects the reduction of the vapor density over very dry soil, which limits the supply of vapor to the overlying air. In fugacity-impedance terms, the effect can be described as an increase in impedance in the surface layer due to the reduction in the fugacity of the herbicide in the dry soil.

In other experiments in Ohio, reported by Taylor et al.,[46] where heptachlor and dieldrin were incorporated to the 7.5-cm depth before corn planting, marked midday maxima were observed on days in which the surface soil remained moist and there was free water evaporation. These diurnal variations were shown to be closely correlated with the evaporation of soil water as measured in an adjacent lysimeter, confirming the importance of the wick effect in maintaining the supply of heptachlor at soil surface, as discussed in Section III.B.2. The changes in daily volatilization rates over the growing season could also be interpreted in terms of the wick effect, the reduction in volatilization after late June, reflecting the decrease in soil water evaporation as crop transpiration became dominant.

After mid July, volatilization rates were greatly reduced, although 85% of the residue still remained in the soil; diurnal variations with noon maxima were also much less prominent, suggesting that the volatilization became controlled by diffusive flow rather than the wick effect. The results of another and later field experiment at Harford, New York,[63] in which trifluralin was applied at 0.7 kg/ha and immediately incorporated to the 7.5-cm depth in a gravelly loam, suggested that diffusion controlled flow was the limiting mechanism throughout the growing season. Even though the soil was moist on all the days when volatilization fluxes were measured, no diurnal variations with marked midday maxima were observed and the rates were very uniform from early May to the end of July, the daily losses ranging between 0.22 and 0.4 g/ha/day. Comparison with the laboratory data obtained by Spencer and Cliath,[19] which indicated that the limiting rate of diffusion controlled flux of trifluralin mixed into Gila silt loam at 14 kg/ha was between 2.5 and 4 g/ha/day after 20 days, reveals an acceptable measure of agreement between the laboratory and field studies when allowance is made for the different application rates and soil conditions.

None of these experiments revealed meaningful correlations between volatilization rates and air or soil temperature. The results confirm the importance of soil moisture conditions as the dominant factor in controlling the volatilization rate of soil-incorporated herbicides. Where soil moisture moves continuously to the surface in response to surface evaporation, the volatilization is controlled by the wick effect and reflects the residue concentration at the soil surface and the rate at which this is replenished by the upward convective flow of the soil moisture. Marked diurnal variations with midday maxima are often observed as the rate of water evaporation changes with insolation or energy advection by the wind.

When the soil remains moist, but there is little water evaporation, volatilization becomes controlled by diffusive flow; both daily and long term volatilization then tend to be slower and more uniform. Variations probably reflect episodes of soil drying, limited water flow, and temperature changes.

Where the surface soil dries to a moisture content less than 2 to 5%, depending upon soil texture, volatilization is reduced to very low values or ceases completely due to reduction of the pesticide vapor density by increased adsorption on the dry soil. This effect is, however, reversible and some volatilization may be observed in the evening and early morning as the surface is wetted by dew accumulation; very rapid and marked increases may occur when soils are wetted by rainfall. Since air and surface soil temperatures may rise very steeply over dry soil that is not cooled by water evaporation, no meaningful relationship can be expected between volatilization and temperature or windspeed.

All these field experiments indicate clearly that the volatilization rates of incorporated herbicides are controlled by soil conditions. Meteorological variables, such as temperature, radiation input, and windspeed, are important insofar as they control the soil conditions, but it is the latter that are critical in controlling the supply of available herbicide at the surface, which in turn controls the volatilization rate.

b. Surface Applications

One of the most conspicuous effects of soil moisture levels on the volatilization of surface residues was observed by Turner et al.[35] who reported the results summarized in Table 5. In this experiment, a spray of a water-based emulsion of chlorpropham was applied to a very dry bare soil. Despite continuous sunshine and a steady wind, the volatilization rate decreased steadily during the first 3 hr as the moisture applied in the spray was evaporated. Despite a 4°C rise in air temperature, the volatilization rate only increased after the soil was moistened by a rain shower at 1500 hr. On the morning of the third day, when the soil moisture had been raised to 18% by an intervening period of rain, the volatilization exceeded the initial rate, but decreased again in the afternoon as the surface soil dried, despite a rise in temperature (Table 5). A similar effect was observed by Glotfelty[50] in a field experiment where trifluralin, heptachlor, and lindane were applied together as a water-based spray of emulsifiable concentrate to a sandy loam soil at Salisbury, Maryland. Although the soil surface was initially moist at the 6 a.m. application, the volatilization rates of all the pesticides decreased as the surface of the soil dried during the day; all increased again as the soil moistened in the evening. The limiting effect of dry soil surfaces was also evident on the second day of the Salisbury experiment, when the volatilization rates of all the pesticides were reduced to their lowest levels as the soil again dried in the early afternoon (Table 6). In contrast, the chlorpropham experiment showed a maximum volatilization in the early afternoon of the third day after the soil had been moistened by rain. In other experiments where trifluralin and dacthal (together with heptachlor and chlordane) were applied to a silt loam soil which did not dry during the day, marked noonday maxima were observed.[26] In later work with surface applications of alachlor, atrazine, and simazine,[53] the alachlor vol-

Table 5
EFFECT OF SOIL MOISTURE AND TEMPERATURE ON THE VOLATILIZATION OF SURFACE-APPLIED CHLORPROPHAM

Day	Sampling period (EDT)	Temperature (°C)	Volatilization rate (g/ha/hr)
1	11—1200		29.4
	12—1300		10.0
	13—1400	20.6	7.6
	14—1500		10.4
	15—1600	16.8	28.7
3	07—0900	19.0	2.0
	09—1100		17.6
	11—1300	22.0	21.9
	13—1500		21.8
	15—1700	23.0	8.9
	17—1800		4.0

From Turner, B. C., Glotfelty, D. E., Taylor, A. W., and Watson, D. R., *Agron. J.*, 70, 933, 1978. With permission.

Table 6
DIURNAL CHANGE IN VOLATILIZATION RATES OF SURFACE-APPLIED TRIFLURALIN, HEPTACHLOR, AND LINDANE FROM SANDY LOAM SOIL

Date and time (EDT)	Volatilization rate (g/ha/hr)			Heptachlor/-trifluralin ratio[a]
	Trifluralin	Heptachlor	Lindane	
June 14				
0900	11.0	57.0	13.0	3.4
1030	7.2	41.0	8.5	3.8
1230	4.3	32.0	6.2	4.9
1515	1.0	21.0	4.0	13.9
1645	5.1	9.2	2.0	1.2
June 15				
0700	5.1	32.0	5.9	4.2
0815	3.6	17.0	3.9	3.1
0930	1.2	8.5	2.3	4.7
1045	0.7	5.8	1.3	5.5
1200	1.1	8.6	2.1	5.1
1315	1.0	9.4	1.9	6.2
1500	0.7	6.9	1.6	6.5
1615	0.7	7.1	1.8	6.7
1830	0.3	5.5	0.8	12.1
1945	0.7	11.0	3.0	10.4
2110	1.3	24.0	7.4	12.2

Note: Application from 0610 to 0640 a.m. June 14. Each time is midpoint of a 1 hr sampling period.

[a] Ratio of volatilization rate per unit weight of residue

atilization showed afternoon minima on the first and second days, when the soil was dry, but the pattern was reversed to give noonday maxima on the fourth day after rain. The volatilization of both triazines was different, with diurnal curves that followed the alachlor on moist days, but did not show reductions over dry soil. This suggests that atrazine volatilization is not limited by soil moisture deficits. This question will be examined below.

Very high volatilization losses of EPTC were measured by Cliath et al.[48] during flood irrigation of an alfalfa field in California. The measured loss of 2.24 kg/ha from water and moist soil over a 52-hr period represented 74% of that applied in the irrigation water.

Measurements of the volatilization of the octyl ester of 2,4-D after application to 10-cm-high wheat in Saskatchewan in May[49] showed a more complex pattern. During the first two days after application, maximum volatilization rates occurred in the early afternoon, despite very low surface soil moisture levels at these times, and a third volatilization peak was observed between 0230 and 0430 a.m. on the morning of the third day, immediately following a rain shower. This anomalous pattern was resolved by determinations of the changes in the residue levels on the soil and crop surfaces, which clearly showed that the maxima of the first two days were due to volatilization from the residues on the plant leaves, which had intercepted 52% of the application; there was little or no contribution from the soil during this time. The morning maximum on the third day was, however, due to the desorption of the soil residues following the rain. These results indicate the complexity that may be encountered in volatilization patterns where surfaces with different characteristics are present at the same time. Other work with insecticides has shown that volatilization from transpiring plant leaves takes place freely during the day, even when the underlying soil surface is dry.[54,56]

These observations all confirm the primary importance of the moisture content of the surface soil layer in controlling volatilization. In the case of surface applications, the effect is clearly due to the reduction in equilibrium vapor density or fugacity of the residue deposit by the increased adsorption in the dry soil. Although there is some evidence that when the soil remains moist, the time-course of the disappearance of surface residues is controlled by "back-diffusion", as discussed in Section III.A.,[26] the moisture effects are generally simpler and more direct than those observed with incorporated residues because the wick effect and long-period diffusion mechanisms are not encountered. As is evident from Table 4, disappearance times of surface residues are much less than with incorporation, except perhaps for compounds which have low vapor pressures and Henry's law coefficients.

3. Comparative Behavior of Different Compounds

While it is evident from Table 4 that surface applications of the more volatile pesticides, such as trifluralin and EPTC, are the most rapidly lost, the previous discussion shows that the volatilization rates are highly dependent upon the soil moisture content and the way in which the pesticides are used; these effects are evident in the trifluralin data in Table 4. Direct comparisons between individual compounds can only be made in field experiments where they are applied together (preferably in the same spray) so that the usage, weather conditions, and sampling schedules are identical. Few results from such experiments are available, but comparisons are possible for a limited number of compounds; these include trifluralin, heptachlor, lindane, and dacthal[25,26,50] and alachlor, atrazine, and simazine.[53] Data are presented in Table 7 comparing volatilization rates of trifluralin and heptachlor during the first 36 hr after a surface application to a moist silt loam as a single spray of water-based emulsifiable concentrate.[50] These results show a close parallel in the behavior of the two compounds. When expressed in terms of the specific volatilization rate (volatilization per unit weight of residue), the heptachlor was being lost between 1.7 and 1.9 times as fast as the trifluralin during the first 12 hr, and between 1.4 and 2.2 times as fast on the second day. The ratios presented in Table 7 suggest that the relative loss of trifluralin was

Table 7
DIURNAL CHANGE IN VOLATILIZATION RATES OF
SURFACE-APPLIED TRIFLURALIN, DACTHAL, AND
HEPTACHLOR FROM MOIST SILT LOAM SOIL

Date and time (EDT)	Volatilization rate (g/ha/hr)			Heptachlor/-trifluralin ratio[a]
	Trifluralin	Heptachlor	Dacthal	
August 8				
0900	38.0	76.0	1.1	1.7
1000	56.0	117.0	2.6	1.7
1100	58.0	117.0	3.3	1.7
1200	33.0	71.0	2.8	1.8
1300	38.0	83.0	4.9	1.8
1400	30.0	65.0	4.8	1.8
1500	27.0	62.0	5.3	1.9
1600	16.0	35.0	3.2	1.8
1700	10.0	21.0	2.1	1.7
1800	9.4	20.0	2.0	1.7
August 9				
1030	4.3	7.3	1.3	1.4
1130	6.8	12.0	3.3	1.5
1230	9.7	20.0	6.0	1.7
1330	9.6	21.0	6.8	1.8
1430	9.7	24.0	7.9	2.1
1530	5.8	15.0	5.2	2.2
1630	4.6	11.0	4.2	2.0
1730	4.2	10.0	4.1	2.0

Note: Application from 0645 to 0730 a.m. August 8. Each time is midpoint of a 1-hr sampling period.

[a] Ratio of volatilization rate per unit weight of residue.

becoming slower during the afternoon of the second day, although soil analyses showed that almost equal amounts of the two compounds remained in the soil after 34 hr. Since 26% of the trifluralin and 29% of the heptachlor initially present in the soil still remained after 10 hr, the ratio of the volatilization rates during this period does not appear to have been restricted by depletion of the amounts of either residue, so that the average factor of 1.75 probably represents a fair comparison of the relative volatilization of the residues in this experiment.

A similar approximate agreement was found in a second experiment[26,50] comparing the volatilization rates of surface residues of trifluralin and lindane for 36 hr after application; here, the initial lindane/trifluralin flux rate was 0.65 and averaged 0.56 over the first 6 hr. The lower ratio probably reflects the lower volatility of the lindane, whose saturation vapor pressure is about half that of trifluralin (Table 1).

The heptachlor/dacthal ratios in the first experiment were less consistent, showing a steady decrease due to the smaller and lower specific volatilization rates of the dacthal residues. This herbicide was applied as a wettable powder, so that the release of the active ingredient may have been slower than that of the heptachlor.

The heptachlor/trifluralin volatilization ratios in Table 6, obtained in a third experiment where the volatilization was inhibited by soil drying,[50] also show a different pattern, with the values of the specific ratios being generally higher than those in Table 7, and tending to rise late in the day. Inspection of the hourly volatilization rates shows that this was because the trifluralin rate remained low while the heptachlor increased as the soil remoistened in the evening. This indicates that the reduction in fugacity of the two materials caused by adsorption on the dry soil was not the same, and that it did not recover at the same rate as the soil rewetted. This result is important because it shows that comparisons of the volatilization rates of two compounds, even under identical experimental conditions, can only be expected to reflect the vapor pressures of the pure materials as long as they are evaporating from surfaces where neither is adsorbed. Where adsorption does take place, as in dry soil, the relative rates of volatilization will be a complex function of the energies of adsorption in addition to the chemical properties of the pure compounds. When soil moisture changes during the experiment, so that adsorption characteristics also change, kinetic effects due to hysteresis between adsorption and desorption isotherms are also likely to be important.

4. Effects of Special Utilization Patterns

The results summarized in Table 4 include data obtained from surface applications of a microencapsulated formulation of chlorpropham and several materials applied on wettable powders (atrazine, simazine, and dacthal); all these show notably lower rates of disappearance than those applied as water-based emulsions.

a. Microencapsulation

The volatilization of chlorpropham from a microencapsulated formulation was compared with that of a conventional emulsion in a field experiment reported by Turner et al.[35] The formulation was composed of a chlorpropham solution contained in 25-μm nylon capsules suspended in water and applied as a water-based spray. The specific volatilization rate of the encapsulated formulation was 12% of the conventional on the first day of the experiment, 26% on the third, and 18% on the eighth; soil residue analysis showed that the 50% disappearance time was about 21 days for the encapsulated formulation and 9 days for the conventional. Comparison of the rate of disappearance with the estimated volatilization loss showed that over a 50-day period, about half the disappearance of the conventional application could be attributed to volatilization. Although the amount of encapsulated material disappearing over the same time was much less, corresponding to the longer "half-life", about one quarter of the loss could still be attributed to volatilization. The rest of both formulations were lost by chemical or biochemical degradation. Although the vapor densities were lower over the encapsulation plot, both showed that same general patterns, with maximum fluxes at midday only when the soil was moist. This similarity suggests that the volatilization was not controlled by direct release from the capsules themselves, but was associated with revolatilization of chlorpropham that had been adsorbed on the soil surfaces as it was slowly released from them. Detailed interpretation of the mechanism is not possible on the basis of the data available, but the results do indicate that where the increased cost can be justified, and there is no reduction in biological activity, encapsulation techniques offer real possibilities for the reduction of the environmental impact of some herbicides. Field evaluations published by Gentner and Danielson[57] showed that the formulation used by Turner et al.[35] was herbicidially effective for a significantly longer period of time than the conventional emulsions.

b. Wettable Powder Formulations

Measurements of volatilization of herbicides applied as water-based sprays of dispersed wettable powders have been made in two field experiments with atrazine and simazine[53] and dacthal.[50] In both cases, there was some evidence that in addition to movement of the herbicides as molecular vapor, there was some erosion and movement of the wettable powder in the air, giving an apparent increase in the measured volatilization rate, particularly under dry soil conditions. In both experiments, the evidence was obtained with a set of air samplers containing plugs of glass fiber filters mounted alongside the regular polyurethane plug vapor samplers. The initial objective of the measurements was the measurement of the amount of particulate-borne herbicide retained by the polyurethane plugs. In the dacthal experiment, the amounts found on the glass fiber filters indicated that about 20 to 30% of the dacthal in the lowest part of the profiles was carried by the particulates, whereas all the other pesticides in the experiment - trifluralin, heptachlor, and chlordane - were entirely in the vapor form. The fraction of dacthal carried by the particulates increased with increasing speed and decreased with height, causing the profile shapes to change in a characteristic way in response to changing wind and soil conditions. In laboratory tests, the same fibers exposed to vapors alone retained very small amounts of residues.

In the second field experiment, both alachlor and triazine herbicides fluxes over moist soil showed diurnal variations with midafternoon maxima, but with dry soil conditions, the triazines again showed a small midafternoon maximum when the alachlor volatilization was inhibited. Data from glass fiber samplers again suggested that these maxima over dry soil were due to wind erosion of the formulation particles.

These results are of particular importance in the interpretation of herbicide "vapor" profiles of wettable powder formulations over dry soils. These residues appear to differ from those applied as emulsions because they remain on the soil surface as thin deposits of very fine powder containing a very high pesticide concentration which does not diffuse into the soil surface as happens with liquid emulsion formulations. When the soil is moist, these particles are probably held on the soil surface by capillary forces, but on drying, their relatively small size and the failure of the capillary forces makes them more subject to removal by wind erosion than the larger soil particles. It is noteworthy that this particle transport has been seen only with compounds of lower volatility, where it may be expected to be most prominent. For compounds of higher volatility with profiles of much higher vapor concentrations, the fraction moving in particulate form would be much smaller and difficult to observe. The possible importance of particulate movement in the long distance transport of slightly volatile compounds should not, however, be discounted.

c. Volatilization from Water Surfaces

The volatilization of herbicides after application to open water surfaces for the control of aquatic vegetation or weed control in a rice paddy presents a case of special environmental interest and importance. The factors controlling the distribution of the herbicide between the air, water, and sediment phases of this system will of course differ from those in soils, with the effects of thermal- and wind-induced mixing and turbulence being important. The fundamental principles relevant to the investigation of such systems have been discussed by Hartley and Graham-Bryce,[58] but no results from field measurements in such systems are currently available in the literature.

VI. AGRONOMIC AND ENVIRONMENTAL SIGNIFICANCE

A. Agronomic Implications

The research summarized above shows that volatilization is a major cause of disappearance of herbicides from target areas, particularly where they are surface applied, and the rate of this loss can exceed that by chemical degradation. Recognition of this fact is of major importance for improved herbicide management. The most unfavorable conditions are surface applications to moist soils. In no-tillage cropping practices, where volatilization losses may well prove to be most serious, applying herbicides only when the soil is dry is likely to be the best preventive measure. Although volatilization losses will certainly occur when the soil is moistened by rain, the loss is likely to be less than those that have been found after direct application to moist soil surfaces where there may be little tendency for the formulation to diffuse downward away from the outer surface soil layer before volatilization begins.

Where choices of chemicals are possible, the selection of less volatile herbicides would be clearly desirable.

Where cultivation of the soil surface is consistent with the soil management practice, as in reduced or minimum tillage systems, incorporation immediately after application, even to a very shallow depth, will greatly reduce the loss of even the more volatile herbicides.

Attempts to prevent herbicide losses by the use of encapsulated formulations have been shown to be effective, but may not be economically feasible since the increased cost may be greater than that of using the larger amounts of herbicide in conventional formulations needed to offset the volatilization loss. The need for special equipment to apply such formulations - which sometimes cause settling and distribution problems in storage and spray tanks - also represents a further increase in cost. The greater environmental impact associated with conventional formulation must also be balanced against the increased risk of movement of encapsulated formulations in surface runoff since these may be detached from soil surfaces with an increased direct impact on water quality in streams or waterways.

B. Environmental Significance

The most important environmental aspect of herbicide volatilization is the potential for the rapid injection of a large fraction of the applied residue into the atmosphere. The further possibility of the rapid transport of the resulting vapor over long distances makes this a distribution pathway of particular importance. The amounts lost in runoff water or on eroding soil are generally small in comparison to potential volatilization losses; the distances over which these move are much smaller, although their concentrations may be locally much higher with greater immediate potential risks to sensitive species of aquatic organisms.

The large distances over which the airborne residues may be carried is offset by the rapid dilution in the atmosphere. This, coupled with the possible degradation by photochemical and oxidative reactions, reduces the risk of acute environmental impacts. These cannot, however, be entirely discounted. As noted elsewhere, the effects of vapor movement on sensitive crops at considerable distances from the point of application have been documented. Reconcentration of vapor by adsorption into rain and fog droplets, with possible redeposition on vegetation has also been observed. While a full assessment of the behavior of herbicide vapor in the atmosphere lies outside the scope of this chapter, it should be noted that a number of the chemical properties that affect herbicide volatilization are also likely to be important in controlling any bioaccumulation that may take place in plant or animal tissue, or in environmental components remote from the point of use. The fugacity approach discussed by Mackay[4] shows great promise for predicting places where the greatest risks may exist for particular compounds. To date, no chronic adverse impacts have been reported for herbicides similar to those caused by the organochlorine insecticides, probably reflecting the generally higher water solubilities and lower chemical stabilities of herbicide compounds. The possibility of the future identification of such effects cannot be entirely discounted as improvements in toxicological techniques are forthcoming. The sensitivity of chemical analyses for the measurement of very low levels of residues in many environmental compartments is approaching the point at which predictions of bioaccumulation can be experimentally examined at all but the lowest levels.

In assessing the environmental impacts of herbicide use it should be recognized that there are also very direct soil and water conservation benefits obtained from their use in minimum tillage or no-tillage farming systems, where the short-term impacts of their use are more than offset by the long-term benefits in the reduction of erosion and improvement in water quality associated with these management practices.

C. Modeling Requirements

Both agronomic effectiveness and environmental impact are very complex processes that will be better understood as our capabilities for predictive modeling increase. At the present time, these are somewhat limited as far as detailed predictions of volatilization events, analogous to runoff events, are concerned. Two kinds of models are available reflecting the macro and micro approaches to the problem. Macroscopic modeling predicts the ultimate distribution of a particular chemical between the large scale phases of the environment — the overall atmosphere, water bodies, and the sediments and soils.[59] The fugacity approach[60] provides a good basis for this, but although it can take into account the effect of differing degradation rates in each phase, it does not contain any description of the factors that control the kinetics of the exchanges between the phases. Volatilization from soil to air is one of these exchange routes. In contrast, micro models of the type presented by Mayer et al.[32] and Jury et al.[8,30,61,62] describe the factors controlling redistribution within the soil. While these models describe the rate at which available herbicide residues are delivered to the surface, which is of major importance in controlling volatilization, they tend to assume fairly constant conditions within the soil for the time required for equilibrium or steady state conditons to develop, and do not predict the effects of rapid changes in soil moisture content and flow. Thus, while they are of great value in predicting the differences in behavior of chemicals under a variety of different management practices, they are limited in their capability for describing the impact of the rapidly varying conditions in the field.

Future improvements in successful predictive models that can provide a basis for practical and rapid management decisions using real time data depend upon the extension of these models into the ground between them — the meso scale. Such models need not be too precise. As is clear from the earlier discussion, the precision with which we can measure volatilization rates is, and may remain, limited, owing to the difficulties and expense of obtaining "real time" data in the field. This is also true for measurements of the amounts of residues present on soils and plants in the field. There is no point in attempting to describe or predict volatilization or other losses with a precision greater than they can be measured. Volatilization losses are also weather dependent and cannot be forecast with more confidence than the weather can be predicted. It therefore appears that the most successful approach to mesoscale modeling is likely to be aimed at predicting the probable consequences of use patterns and management decisions, rather than attempting to make detailed predictions of the amounts of pesticides that will be lost from soil to air from specific regions or land areas.

VII. SURVEY AND CONCLUSIONS

A. Theoretical Understanding

The principal weaknesses in our understanding of the volatilization process lie in the theory by which measurements of profile gradients are used to calculate flux rates. There are two weaknesses; both lie in the conceptual basis of the underlying micrometeorological theory.

The first is our understanding of the nature of the laminar flow layer that is conceived to exist on the surface of soil and plant surfaces. The idea of the existence of this layer is derived from aerodynamic theory which postulates it as a smooth layer over an aeroplane wing or other surface exposed to flowing air through this layer. The drag forces are transmitted to the surface of the soil through this layer. Surface smoothness is a key element in this concept. The surfaces of soils and plants are, however, not smooth on the cellular or micropore scale and it seems possible that turbulent mixing may penetrate within their pore structures so that laminar flow may only exist for short and transient periods, punctuated by rapid exchanges or microbursts of external air. The gas and momentum exchange may take place more by a type of pumping action rather than diffusive flow. The question as to which concept is more valid is important in modeling the release of vapor and in locating the physical boundary at which the dispersion of the pesticide vapor becomes passively controlled by turbulent mixing, rather than diffusive molecular flow. This is important for improving the models because it defines the level within the soil where the transition is believed to take place. It will be particularly important in relating the rate of vapor formation to the soil water content and the extent to which free water surfaces extend to the soil surface where there is immediate turbulent dilution.

The second weakness lies in the assumptions made in using the "similarity principle" which equates the diffusivity coefficients of pesticide vapor and momentum (or heat transfer) at any point in the atmospheric boundary layer. Detailed comparisons of profiles of different compounds flowing through the same layer at the same time show differences which suggest that the diffusivity coefficients for different compounds are not the same. This throws grave doubt on the assumption that they can be identified with the momentum coefficient, an assumption whose main justification is that it is convenient to make it. Comparison of calculated loss rates with measured residue disappearances in the experiments where we may expect to find reasonable agreement when the sampling uncertainties are discounted, also suggests that the flux rates calculated using the similarity principle may sometimes underestimate the actual values. The resolution of this discrepancy is an old problem in micrometeorology, but one which merits more detailed examination. Using sensitive analytical techniques we can follow the dilution of pesticides and other organic molecules to concentrations which are not accessible in profiles of water vapor, carbon dioxide, and ammonia that are superimposed upon atmospheric background levels and do not extrapolate to zero. While the resolution of this issue may not lead to improved confidence in the calculated values by validation of the similarity principle, it may be expected to improve our knowledge of the atmospheric conditions under which it breaks down and thus evaluate the uncertainties in our present experimental methods.

Fundamental weaknesses also exist in our knowledge of the values of basic parameters used to characterize the chemical properties of pesticide compounds used in modeling. The values for the Henry's law coefficients are a clear example. All the values presented in Table 2 are constructed from measurements of saturation vapor pressures of pure compounds and their saturation solubilities in water. Few direct measurements of this coefficient, or the way in which it varies with concentration and vapor pressure, are available. Similar deficiencies exist with soil adsorption coefficients which are measured in terms of disappearance from water solution without any allowance for the hysteresis effects in desorption. Since these data are of fundamental importance as parameters in modeling predictions, our ignorance of the validity of these numbers again represents a major weakness that can only be repaired by continued research efforts on selected compounds where a considerable body of field and laboratory data already exists.

B. Measurements

One of the principle difficulties that limits the amount of data available from field experiments on volatility is the number of samples that must be taken to obtain the profiles necessary for flux calculations. There is a clear need for the development of simplified procedures for making reliable flux measurements with fewer pesticide analyses and less meteorological data. Methods using horizontal, rather than vertical, flux measurements would permit much simpler and more flexible field experimentation. Further work on the significance of particle transport and identification of the conditions under which wind erosion of surface-applied powder formulations is important, and would also be a contribution to our understanding of the atmospheric transport of some of the more important and less volatile materials, particularly in relation to our knowledge of the conditions under which these applications may act as sources for long distance transport of materials such as triazines.

C. Significance

Although herbicide losses by volatilization vary a great deal depending upon soil conditions and the way they are used, this is often the principal pathway by which residues are lost from target areas. Such losses can exceed 90% of the application within 48 hr or less when residues of more volatile compounds are exposed on moist soil or plant surfaces.

The principal factors controlling the rate of volatilization include the chemical properties of the particular herbicide, the way in which it is used, and the soil moisture supply. Volatilization is greatly reduced by incorporation into the soil when the rate becomes dependent upon convective or diffusive flow of the chemical to the soil surface. Both these mechanisms are very sensitive to soil moisture conditions. Convective flow is dominant where herbicide is carried to the surface by upward movement of soil moisture. Diffusion controls the supply of herbicide to the surface in moist soil where there is no upward moisture flow. Changes in moisture flow during the day control the diurnal patterns of volatilization. Most herbicides are strongly adsorbed by dry soil so that volatilization of both incorporated and surface-exposed residues ceases almost completely from dry soil surfaces.

Volatilization rates can be reduced by the use of special formulations, such as microencapsulation, but the use of such formulations is limited because of increased cost and technical problems that require modifications of regular application equipment. The loss of effectiveness due to volatilization is more usually offset by using larger application rates than would otherwise be necessary. This results in agronomic effectiveness being maintained at the expense of a greater environmental impact.

Volatilization is a pathway of major importance for the dispersal of herbicide (and insecticide) residues to the general environment. Large fractions of the amounts applied can be rapidly lost and readily transported considerable distances from target areas. Although the speed and ease of these losses are to some degree offset by rapid dilution in the atmosphere, removal and reconcentration of the vapor by adsorption in rain and fog droplets, with consequent redeposition, is a known pathway that may represent the first step in the bioaccumulation of those herbicides (or their degradation products) that are sufficiently stable and have appropriate chemical properties. Where sensitive organisms are present in the environmental phase where reconcentration occurs, the possibilities of chronic low-level toxic effects should not be discounted without evaluation.

REFERENCES

1. **Wauchope, R. D.,** The pesticide content of surface water draining from agricultural fields - a review, *J. Environ. Qual.,* 7, 459, 1978.
2. **Monteith, J. L.,** *Principles of Environmental Physics,* American Elsevier, New York, 1973, chap. 2.
3. **Hartley, G. S. and Graham-Bryce, I. J.,** *Physical Principles of Pesticide Behavior,* Vol. 1, Academic Press, New York, 1980, chap. 6.
4. **Mackay, D.,** Air/water exchange coefficients, in *Environmental Exposure from Chemicals,* Vol. 1, Neely, W. B. and Blau, G. E., Eds., CRC Press, Boca Raton, Florida, 1985, chap. 5.
5. **Lewis, G. N.,** The law of physico-chemical change, *Proc. Am. Acad.,* 37, 49, 1901.
6. **Mackay, D.,** Calculating fugacity, *Environ. Sci. Tech.,* 5, 1006, 1981.
7. **Hartley, G. S. and Graham-Bryce, I. J.,** *Physical Principles of Pesticide Behavior,* Vol. 2, Academic Press, New York, 1980, 896 (Appendix 4).
8. **Jury, W. A., Spencer, W. F., and Farmer, W. A.,** Use of models for assessing relative volatility, mobility, and persistence of pesticides and other trace organics in soil systems, in *Hazard Assessment of Chemicals: Recent Developments,* Vol. 2, Saxena, J., Ed., Academic Press, N.Y., 1983, 1.
9. **Beste, C. E. and Humburg, N. E.,** *Herbicide Handbook of the Weed Science Society of America,* 5th ed., Weed Science Society, Champaign, Illinois, 1983.
10. **Plimmer, J. R.,** Volatility, in *Herbicides: Chemistry, Degradation and Mode of Action,* Vol. 2, Kearney, P. C. and Kaufman, D., Eds., Dekker, New York, 1976, 891.

11. **Hamaker, J. W. and Kerlinger, H. O.,** Vapor pressure of pesticides, in *Pesticidal Formulations Research: Advances in Chemistry Series No. 86,* Gould, R. F., Ed., American Chemical Society, Washington, D.C., 1969, 39.

12. **Spencer, W. F. and Cliath, M. M.,** Vapor density of dieldrin, *Environ. Sci. Tech.,* 3, 670, 1969.

13. **Hartley, G. S. and Graham-Bryce, I. J.,** *Physical Principles of Pesticide Behavior,* Vol. 2, Academic Press, New York, 1980, chap. 11.

14. **Spencer, W. F.,** Distribution of pesticides between soil, water and air, in *Pesticides in the Soil: Ecology, Degradation and Movement,* Michigan State University Press, East Lansing, 1970, 120.

15. **Spencer, W. F.,** Volatilization of pesticide residues, Technical Seminar, Fate of Pesticides in the Environment, University of California Agricultural Experiment Station Special Publ., 1985.

16. **Spencer, W. F., Cliath, M. M., and Farmer, W. J.,** Vapor density of soil-applied dieldrin as related to soil-water content, temperature and dieldrin concentration, *Soil Sci. Soc. Am. Proc.,* 33, 509, 1969.

17. **Spencer, W. F. and Cliath, M. M.,** Desorption of lindane from soil as related to vapor density, *Soil Sci. Soc. Am. Proc.,* 34, 574, 1970.

18. **Spencer, W. F. and Cliath, M. M.,** Volatility of DDT and related compounds, *J. Agric. Food Chem.,* 20, 645, 1972.

19. **Spencer, W. F. and Cliath, M. M.,** Factors affecting vapor loss of trifluralin from soil, *J. Agric. Food Chem.,* 22, 987, 1974.

20. **Fang, S. C., Theisen, P., and Freed, V. H.,** Effects of water evaporation, temperature and rates of application on the retention of ethyl-*N, N*-di-*n*-propylthiolcarbamate in various soils, *Weeds,* 9, 569, 1961.

21. **Deming, J. M.,** Determination of volatility losses of C14-CDAA from soil surfaces, *Weeds,* 11, 91, 1963.

22. **Kearney, P. C., Harris, C. I., Kaufman, D. D., and Sheets, T. J.,** Behavior and fate of chlorinated aliphatic acids in soils, *Adv. Pest Control Res.,* 6, 1, 1965.

23. **Gray, R. A. and Weierich, A. J.,** Factors affecting the vapor loss of EPTC from soils, *Weeds,* 13, 141, 1965.

24. **Parochetti, J. V. and Warren, G. F.,** Vapor losses of IPC and CIPC, *Weeds,* 14, 281, 1966.

25. **Harper, L. A., White, A. W., Bruce, R. R., Thomas, A. W., and Leonard, R. A.,** Soil and microclimate effects on trifluralin volatilization, *J. Environ. Qual.,* 5, 236, 1976.

26. **Glotfelty, D. E., Taylor, A. W., Turner, B. C., and Zoller, W. H.,** Volatilization of surface-applied pesticides from fallow soil, *J. Agric. Food Chem.,* 32, 638, 1984.

27. **Nash, R. G.,** Comparative volatilization and dissipation rates of several pesticides from soil, *J. Agric. Food Chem.,* 31, 210, 1983.

28. **Acree, F., Beroza, M., and Bowman, M. C.,** Codistillation of DDT with water, *J. Agric. Food Chem.,* 11, 278, 1963.

29. **Igue, K., Farmer, W. J., Spencer, W. F., and Martin, J. P.,** Volatility of organochlorine insecticides from soil. II. Effect of relative humidity and soil water content on dieldrin volatility, *Soil Sci. Soc. Am. Proc.,* 36, 447, 1972.

30. **Jury, W. A., Spencer, W. F., and Farmer, W. J.,** Behavior assessment model for trace organics in soil. I. Model description, *J. Environ. Qual.,* 12, 558, 1983.

31. **Taylor, A. W.,** Post-application volatilization of pesticides under field conditions, *J. Air Pollut. Control Assoc.,* 28, 922, 1978.

32. **Mayer, R., Letey, J., and Farmer, W. J.,** Models for predicting volatilization of soil-incorporated pesticides, *Soil Sci. Soc. Am. Proc.,* 38, 563, 1974.

33. **Farmer, W. J., Igue, K., Spencer, W. F., and Martin, J. P.,** Volatility of organochlorine insectides from soil. I. Effect of concentration, temperature, air flow rate and vapor pressure, *Soil Sci. Soc. Am. Proc.,* 36, 443, 1972.

34. **Jury, W. A., Grover, R., Spencer, W. F., and Farmer, W. J.,** Modeling losses of soil-incorporated triallate, *Soil Sci. Soc. Am. J.,* 44, 445, 1980.

35. **Turner, B. C., Glotfelty, D. E., Taylor, A. W., and Watson, D. R.,** Volatilization of microencapsulated and conventionally applied chlorpropham in the field, *Agron. J.,* 70, 933, 1978.

36. **Spencer, W. F., Farmer, W. J., and Cliath, M. M.,** Pesticide volatilization, *Residue Rev.,* 49, 1, 1973.

37. **Spencer, W. F. and Cliath, M. M.,** Pesticide volatilization as related to water loss from soil, *J. Environ. Qual.,* 2, 284, 1973.

38. **Hartley, G. S. and Graham-Bryce, I. J.,** *Physical Principles of Pesticide Behavior,* Vol. 1, Academic Press, New York, 1980, chap. 5.

39. **Hartley, G. S.** Evaporation of pesticides, in *Pesticidal Formulations Research: Physical and Colloidal Chemical Aspects,* Advanced Chemistry Ser. 86, American Chemical Society, Washington, D.C., 1969, 115.

40. **Nash, R. G. and Beall, M. L.,** A micro-agroeco-system to monitor the environmental fate of pesticides, in *Terrestrial Microcosms and Environmental Chemistry,* NSF/RA 79-0026, Witt, J. M., Gillet, J. W., and Mayer, J., Eds., National Science Foundation, Washington, D.C., 1979, 86.

41. **Taylor, A. W., Freeman, H. P., and Edwards, W. M.,** Sample variability and measurement of dieldrin content of a soil in the field, *J. Agric. Food Chem.,* 19, 832, 1971.
42. **Taylor, A. W., Caro, J. H., Freeman, H. P., and Turner, B. C.,** Sampling variance in measurements of trifluralin disappearance from a field soil, in *Trace Residue Analysis: Chemometric Estimations of Amount, Uncertainty and Error,* Am. Chem. Soc. Symp. Ser. No. 284, Kurtz, D. A., Ed., American Chemical Society, Washington, D.C., 1984, 25.
43. **Parmele, L. H., Lemon, E. R., and Taylor, A. W.,** Micrometeorological measurement of pesticide vapor flux from bare soil and corn under field conditions, *Water, Air and Soil Pollut.,* 1, 433, 1972.
44. **Lemon, E.,** Gaseous exchange in crop stands, in *Physiological Aspects of Crop Yield,* Eastin, J. D., Haskins, F. A., Sullivan, C. Y., and Van Bavel, C. H. M., Eds., American Society of Agronomy, Madison, Wisconsin, 1969.
45. **Monteith, J. L.,** *Principles of Environmental Physics,* American Elsevier, New York, 1973, chap. 6, 7, and 12.
46. **Taylor, A. W., Glotfelty, D. E., Glass, B. L., Freeman, H. P., and Edwards, W. M.,** Volatilization of dieldrin and heptachlor from a maize field, *J. Agric. Food Chem.,* 24, 625, 1976.
47. **Rose, C. W.,** *Agricultural Physics,* Pergamon Pres, New York, 1966, chap. 2 and 3.
48. **Cliath, M. M., Spencer, W. F., Farmer, W. J., Shoup, T. D., and Grover, R.,** Volatilization of S-ethyl*N,N,*dipropylthiocarbamate from water and wet soil during and after flood irrigation of an alfalfa field, *J. Agric. Food Chem.,* 28, 610, 1980.
49. **Grover, R., Shewchuk, S. R., Cessna, A. J., Smith, A. E., and Hunter, J. H.,** Fate of 2,4-D iso-octyl ester after application to a wheat field, *J. Environ. Qual.,* 14, 203, 1985.
50. **Glotfelty, D. E.,** Atmospheric dispersion of pesticides from treated fields, Ph.D. thesis, University of Maryland, College Park, 1981.
51. **Turner, B. C. and Glotfelty, D. E.,** Field air sampling of pesticide vapors with polyurethane foam, *Anal. Chem.,* 7, 49, 1977.
52. **Bidleman, T. F.,** High volume collection of organic vapors using solid adsorbents, in *Trace Analysis,* Vol. 4, Lawrence, J. F., Ed., Academic Press, New York, 1985, 51.
53. **Glotfelty, D. E., Leech, M. M., Jersey, J., and Taylor, A. W.,** Volatilization and wind erosion of soil-surface applied atrazine, simazine, alachlor and toxaphene, 1986, in press.
54. **Taylor, A. W., Glotfelty, D. E., Turner, B. C., Silver, R. E., Freeman, H. P., and Weiss, A.,** Volatilization of dieldrin and heptachlor residues from field vegetation, *J. Agric. Food Chem.,* 25, 542, 1977.
55. **White, A. W., Harper, L. A., Leonard, R. A., and Turnbull, J. W.,** Trifluralin volatilization losses from a soybean field, *J. Environ. Qual.,* 6, 105, 1977.
56. **Willis, G. H., McDowell, L. L., Harper, L. A., Southwick, L. M., and Smith, S.,** Seasonal disappearance and volatilization of toxaphene and DDT from a cotton field, *J. Environ. Qual.,* 12, 80, 1983.
57. **Gentner, W. A. and Danielson, L. L.,** The influence of micro-encapsulation on the herbicidal performance of chlorpropham, in *Proc. Controlled Release Symp.,* Akron, Ohio, 1976, 726.
58. **Hartley, G. S. and Graham-Bryce, I. J.,** *Physical Principles of Pesticide Behavior: The Dynamics of Applied Pesticide in the Local Environment in Relation to Biological Responses,* Vol.1, Academic Press, New York, 1980, chap. 3 and 7.
59. **Bomberger, D. C., Gwinn, J. L., Mabey, W. R., Tuse, D., and Tsong Wen Chou,** Environmental fate and transport at the terrestrial-atmospheric interface, in *Fate of Chemicals in the Environment,* Swann, R. L. and Eschenroder, A., Eds., American Chemical Society, Washington, D.C., 1983, 197.
60. **Mackay, D., Paterson, S., and Joy, M.,** Application of fugacity models to the estimation of chemical distribution and persistence in the environment, in *Fate of Chemicals in the Environment,* Swann, R. L. and Eschenroder, A., Eds., American Chemical Society, Washington, D.C., 1983, 175.
61. **Jury, W. A., Farmer, W. J., and Spencer, W. F.,** Behavior assessment model for trace organics in soil. II. Chemical classification and parameter sensitivity, *J. Environ. Qual.,* 13, 567, 1984.
62. **Jury, W. A., Spencer, W. F., and Farmer, W. J.,** Behavior assessment model for trace organics in soil. III. Application of screening model, *J. Environ. Qual.,* 13, 573, 1984; **Jury, W. A., Spencer, W. F., and Farmer, W. J.** Behavior assessment model for trace organics in soil. IV. Review of experimental evidence, *J. Environ. Qual.,* 13, 580, 1984.
63. **Taylor, A. W.,** Unpublished data, Soil Nitrogen and Environmental Chemistry Laboratory, Agricultural Research Service, U. S. Department of Agriculture, Beltsville, Maryland.

Chapter 5

DISSIPATION FROM SOIL

Ralph G. Nash

TABLE OF CONTENTS

I. INTRODUCTION

Herbicides dissipate from soil, as do all other organic compounds so far as is known, but dissipation rates vary tremendously among compounds, soils, and climate. The literature on herbicide dissipation has become extensive and continues to increase rapidly. Therefore, a comprehensive review was not conducted, but rather a general view toward understanding of herbicide dissipation per se. Emphasis is placed on herbicide dissipation under field conditions because the field is the real world of concern. Laboratory data helps us understand processes of herbicide dissipation in the field, but direct extrapolation to the field is not usually warranted.

This chapter characterizes the dissipation process and will demonstrate that a conceptual model of herbicide dissipation has been developed based upon observed results. The concept can be presented pictorially, but becomes much more useful when described in a manner that explains the rates, mechanisms, and site of herbicide dissipation at any given time. Describing herbicide dissipation ultimately has been reduced to mathematical equations. The vast amount of data indicates that most data sets on herbicide dissipation can be described by a simple exponential equation corresponding to first-order kinetics. For those data sets that do not follow a simple exponential equation, more complicated equations can and have been developed.

Presently, considerable effort is being directed toward developing general mathematical models to predict herbicide dissipation under field conditions. These range from simple models using a few physicochemical characteristics of the herbicide, to complex models incorporating the physicochemical characteristics of the herbicide and soil and field environment. The simpler models lack precision because most factors affecting dissipation are not included in the model, and the more complex models often require parameters, which affect dissipation, but may not be available. Probably no single model (at least in the immediate future) will be capable of giving acceptable estimations of herbicide dissipation. Rather, an array of models will be necessary to give a range of estimates for the dissipation of a herbicide under particular conditions. It is with these thoughts that this chapter is written.

II. CONCEPTUAL AND OBSERVED CONSIDERATIONS

Weber et al.[1] have conceptualized the fate of herbicides in soils (Figure 1). To depict the dissipation process, Edward[2] proposed a series of line segments, which was modified by Nash,[3] to describe pesticides applied to a field (Figure 2). Initially, the highest loss rates occur during application. Under certain circumstances, application losses may be large, and considerable engineering effort has been expended to limit these losses. Losses during application occur primarily over a short period, perhaps 5 min, but may take as much as 1 hr.[4] Second, volatilization loss apparently occurs for almost all herbicides, primarily because of their distribution over a large surface area or volume, and may be considerable. Volatilization losses are governed by the herbicide vapor pressure, method of application (surface-applied vs. soil-incorporated), formulation, temperature, soil moisture and organic content, relative humidity, wind velocity, and smoothness or roughness of the soil surface. Volatilization losses may be brief for herbicides that degrade rapidly or prolonged for the more persistent herbicides.[5] Most data indicate that volatilization loss rates are reduced greatly after a few hours, or a few days at the most, even for the most persistent herbicides; however, losses through volatilization apparently continue for as long as the pesticide remains.[5-11]

Following volatilization, soil penetration, soil adsorption, or soil leaching, or all to some degree, become dominant factors that govern the herbicide dissipation from soil. Finally, degradation (usually biological, but sometimes partially chemical) of the pesticide become paramount in the disappearance of most herbicides.

FATE OF HERBICIDES (HB)

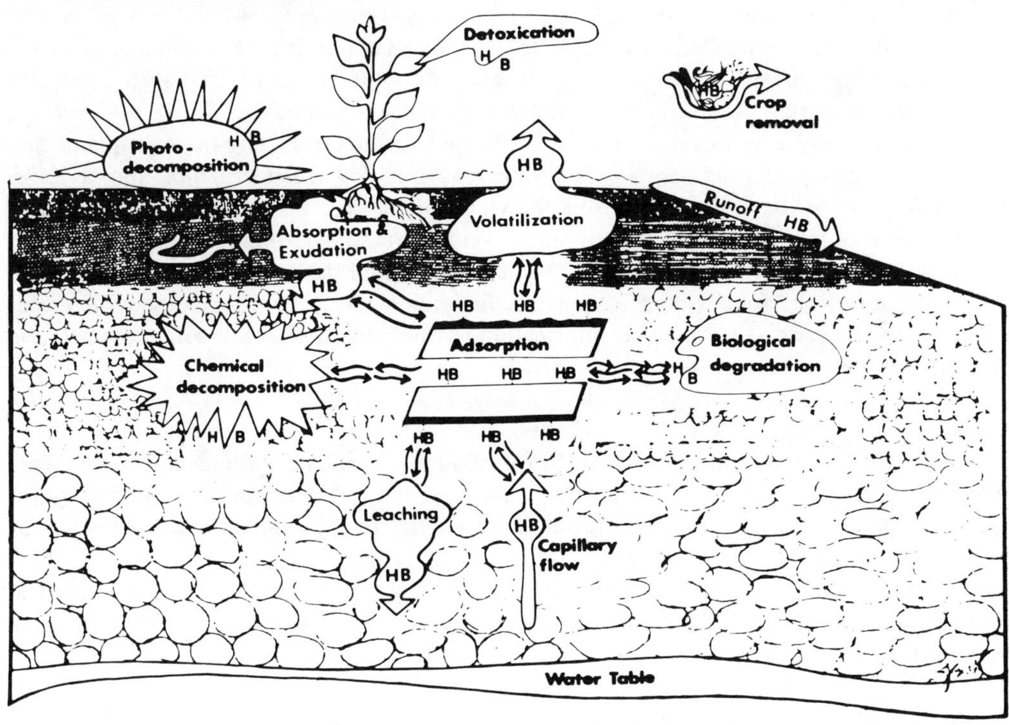

FIGURE 1.   Processes influencing the behavior and fate of herbicides in the environment. Degradation processes are characterized by the splitting of the herbicide (HB) molecule. Transfer processes are characterized by the herbicide (HB) molecules remaining in tact. (From Weber, J. B., Monaco, T. J., and Worsham, A. D., *Weeds Today*, 4, 16, 1973. With permission.)

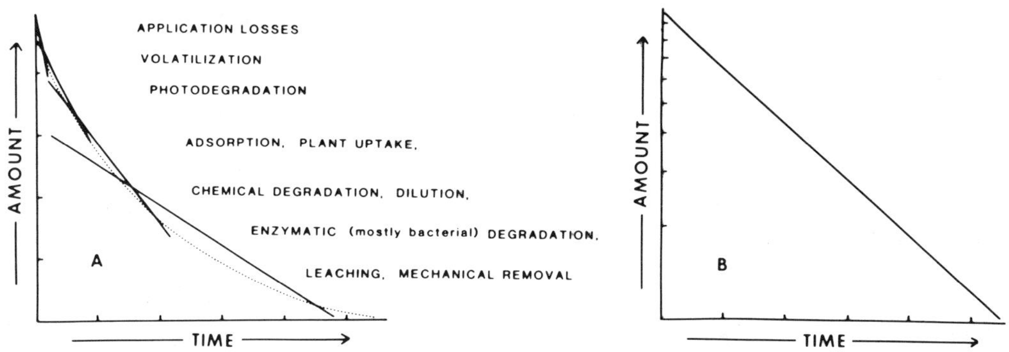

FIGURE 2.   Pesticide losses (dissipation) from soil: (A) arithmetic amount with time and (B) logarithmic amount with time.[3] (From [A] Edwards, C. A., *Residue Rev.*, 13, 83, 1966. With permission.)

A. Distribution

1. During Application

Obviously, if the soil is the target, most of the applied herbicide will reach the soil, except possibly under some aerial application conditions. If a volatile herbicide is surface-applied, considerable amounts may be lost in a short period of time, unless it is immediately incorporated into the rhizosphere.[12] Many herbicides, however, are applied directly to plants, and even then some herbicide usually reaches the soil. Therefore, during foliar application, the herbicide will be distributed primarly among the foliage (target) and soil and air (non-target). The distribution will depend upon the amount intercepted by the foliage or soil surface trash and, as indicated previously, the method of application, formulation, weather, and possibly the physicochemical characteristics of the herbicide itself. Table 1 outlines some of the properties and variables that affect herbicide distribution.

There is no easy way to determine the herbicide distribution among the soil, plant, and air without making measurements. The major reason is that the distribution site, system, and conditions are specific. Usually soil and plant samples are taken or several catchments are placed randomly on the soil surface during application to determine target and off-target herbicide amount. Recently, Willis et al.[13] determined initial insecticide load on cotton from equations developed from cotton insecticide dissipation rates. This would seem a useful technique wherever dissipation-type information is gathered on plants or soil and where the harvested foliage is from a known field area.

2. Subsequent to Application

Researchers have determined herbicide distribution in various environmental components in chamber studies (Table 2).[12,14-19] A significant portion of many herbicides, whether foliar- or soil-applied, ultimately migrates to the atmosphere. The exception is when herbicides were incorporated into the rhizosphere.

The three major recipients of field-applied herbicides are the soil, crop, and air. The atmosphere is not a target, but rather is a sink into which some of the herbicide in or on soil dissipates.[20]

B. Partitioning (Partition Coefficients - K)

Partitioning of a herbicide in soil is the separating and subsequent equilibration of the herbicide into the several soil components: soil particulates (both mineral and organic), soil water, soil air, soil biota, and atmosphere above the soil. Partitioning among the components can be defined by partition coefficients:

$$K_d = \frac{P_S}{P_W} \qquad K_s = \frac{P_A}{P_S} \qquad K_h = \frac{P_A}{P_W} \qquad K_{bcf} = \frac{P_B}{P_S}$$

where P = pesticide concentration in a matrix, S = soil, W = water, A = air and atmosphere, and B = biota. For most herbicides (especially hydrophobic), the separation and equilibration process is rapid, relative to losses from each component. Therefore, time is usually ignored and thermodynamic principles are applicable rather than kinetic principles. K_d (the adsorption coefficient) has been discussed thoroughly elsewhere.[21,22] K_s probably cannot be separated from K_h, and K_h is defined by Henry's law constant. K_{bcf} in soil has not been studied per se, although considerable literature exists on soil fauna and flora herbicide uptake, and plants must take up the herbicide if the herbicide is to be effective. In soil, however, K_{bcf} probably can be treated as a component of the organic matter without serious error. A more thorough discussion of the theoretical aspects of herbicide partitioning in soil can be found in Hartley and Graham-Bryce.[23]

C. Herbicide Dissipation from Soil

The relative persistence of a few important herbicides is shown in Figure 3. The values shown represent a range for 90% loss of the herbicide based upon several field and laboratory data sets. Recently, Jury et al.[24,25] classified the persistence of many of the same herbicides, based upon certain standard properties of the herbicide and soil environment, but with the herbicide placed at a 10-cm soil depth (Table 3). The values given in Table 3 are in good agreement with that shown in Figure 3 and those given in Table 4 for the rhizosphere. The

Table 1
OUTLINE OF THE PROPERTIES AND VARIABLES THAT INFLUENCE HERBICIDE CONTENT ON OR IN SOIL

A. Application
1. Amount per area or volume
2. Formulation
3. Method or placement
4. Plant cover or mulch interception

B. Climatic parameters

1. Average daily temperature
2. Average daily wind run
3. Average daily relative humidity
4. Average daily solar radiation
5. Daily rainfall — time, intensity, duration, and amount
6. Number of rain incidences
7. Subsequent herbicide washoff from plants or plant residue

C. Soil parameters and behavior
1. Type
2. Organic matter content
3. Moisture content
4. Soil pH
5. History of similar pesticide class application(s)

D. Pesticide physical, chemical, and biological properties
1. Molecular weight
2. Transition temperatures
    a. Melting point
    b. Boiling point
    c. Decomposition
3. Vapor pressure
4. Water solubility
5. Partition coefficients (K)
    a. Octanol/water (K_{ow})
    b. Adsorption
        i. Soil (K_d)
        ii. Organic carbon (K_{oc})
    c. Henry's law (K_h)
    d. Bioconcentration factor (K_{bcf})
6. Disassociation
    a. Acid (pK_a)
    b. Base (pK_b)
7. Rate reactions (k)
    a. Volatilization (k_a)
    b. Photolysis (k_p)
    c. Hydrolysis (k_h)
    d. Biodegradation (k_b)
    e. Oxidation (k_o)
    f. Reduction (k_r)
    g. Leaching (k_l)
    h. Diffusion (k_f)
    i. Desorption (k_d)
    j. Bioconcentration (k_{bcf})
    k. Bioelimination (k_e)
    l. Complexation (k_c)

Table 2
HERBICIDE DISTRIBUTION AT END OF EXPERIMENT IN TERRESTRIAL CHAMBERS

Herbicide	Applied to	Distribution, % of applied						
		Soil	Plant	Air	Leachate	Animal	Total	Ref.
2,4-D B E	Ryegrass	15	27	19	—	0.001	61	17
Trifluralin	Fallow surface soil	7	—	60	—	—	67	14
Butylate EC[a]	Rhizosphere	18	5	—	—	—	23[b]	12
Butylate ME[a]	Rhizosphere	69	5	—	—	—	74[b]	12
Bromacil[c]	Various plant spp.	34	34	32	—	0.3	69	19
Trifluralin[c]	Various plant spp.	15	21	62	—	2	72	19
Simazine[c]	Various plant spp.	35	43	21	—	0.8	79	19
2,4,5-T I0 E[c]	Various plant spp.	35	52	12	—	0.7	91	19
PCNB	Seeds in rhizosphere	59	16	24[d]	0.3	0.06	99	18
PCP	Seeds in rhizosphere	65	9	25[d]	0.03	0.3	99	18

[a] EC = emulsifiable concentrate and ME = microencapsulated formulations.
[b] A significant portion of these did go to the air, but degradation most likely resulted in much of that not accounted for.
[c] Based on ^{14}C residues.
[d] Includes estimated aerial losses, which in some cases is questionable.

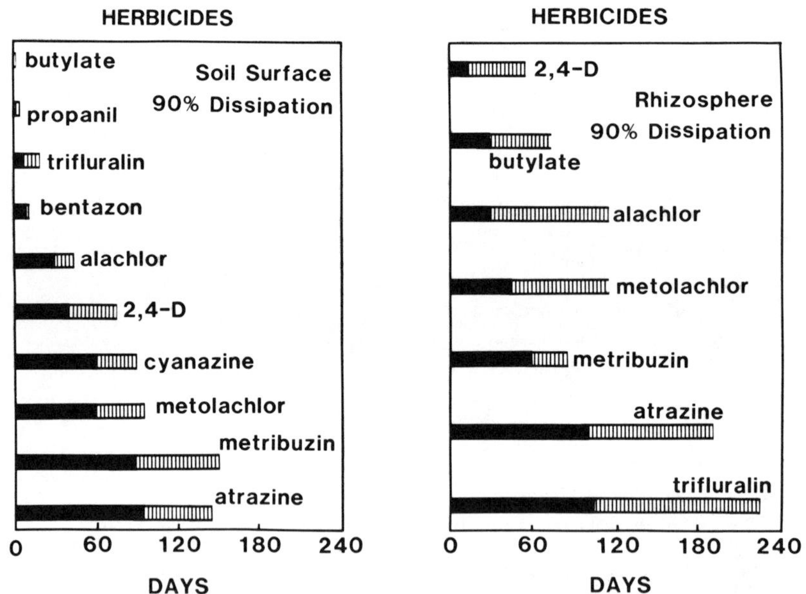

FIGURE 3.   Relative persistence of several important herbicides. The cross-hatched portion of the bars represents typical time ranges for 90% dissipation.

k_s values given in Table 4 refer to the pseudo first-order constant defined by Equation 1, given in Section II.C.1., whereas the k values in Table 3 include other properties in addition to the first-order rate constant. All k values given in this chapter are on a per-day basis.

1. Conceptual and Observed

Hamaker, Leistra, and Soulas have studied the theoretical rates of pesticide degration and decline in soils, and the latter two have developed mathematical models.[26-28] Hamaker

Table 3
DISSIPATION RATES DEFINED IN
TERMS OF EFFECTIVE k_s VALUES
(PER DAY) AT A 10-cm DEPTH
WITH A 0.3 m³ WATER PER m³
SOIL AND NO WATER
EVAPORATION

Herbicide	k_s
Class 1. Highly persistent (k_s = <0.007)	
Bromacil	0.0020
Diuron	0.0021
Monuron	0.0041
Trifluralin	0.0066
Class 2. Moderately persistent (k_s = >0.007 — <0.023)	
Atrazine	0.0098
Napropamide	0.0099
Prometryn	0.0116
Triallate	0.0077
Class 3. Moderately short lived (k_s = >0.023 — <0.046)	
2,4-D	0.0462
EPTC	0.0257
Class 4. Short lived (k_s = >0.046 — <0.139)	
Class 5. Very short lived (k_s = >0.139)	
Ethylene dibromide	0.204
Methyl bromide	1.386

Adapted from Jury, W. A., Spencer, W. F., and Farmer, W. J., *J. Environ. Qual.*, 13, 573, 1984.

suggests that the rate laws are of two basic types: rate = kc^n and rate = $k_1c/(k_2 + c)$, where c is concentration, n is the order of equation, k is the rate constant, k_1 is a maximum rate approached with increasing concentration, and k_2 is a pseudoequilibrium constant.[26] Hamaker found that the former (a power-rate model) usually was the most useful for estimating pesticide dissipation rates, perhaps because not enough observations are made, and that too many complex parameters affect dissipation of pesticides from soil. The latter (a hyperbolic rate model) depends upon pesticide concentration.

The series of straight lines in Figure 2A can be reconstructed as a continuous curve (dashed). Further, using the dashed curve, if the logarithm of the herbicide amount is plotted vs. time, a linear relationship is often possible (Figure 2B). Because this relationship fits first-order kinetics, it is usually referred to as pseudo first-order kinetics. The dissipation of most herbicides from soils is nearly always pseudo first-order or exponential, as has been demonstrated by the data contained in "Environmental Chemical Dissipation File" and the numerous examples presented in this chapter.[29]

Even when dissipation is primarily microbial, herbicide dissipation usually follows pseudo first-order kinetics. Ideally, when dissipation results primarily from biological adaptive

Table 4
SELECTED RATE (k$_s$) VALUES (PER DAY) FOR DISSIPATION OF HERBICIDES FROM SOILS, PRIMARILY FIELD

Pesticide	Dissipation rate (k$_s$)		Soil type	Experimental	Ref.
	Surface	Rhizosphere			
Amides					
Alachlor		0.063—0.077	sl Ustollic Haplargid	Field	40
		0.062	Collington fsl	Field	41
Bensulide		0.014—0.038	Hidalgo sl	Field	42
Metolachlor	0.015—0.033		l	Field	43
Napropamide		0.007—0.055	sl	Field	32
Pronamide		0.0095	sl	Field	44
Propanil		>0.025	Melfort cl	Field	45
Benzoic Acids					
Dicamba		0.0011	Regina heavy clay	Lab	46
		0.019—0.25	Asquith sl	Lab	46
		0.022	Quachita Highlands Forest	Lab	47
		0.023	Cross Timbers Forest	Lab	47
		0.034	Quachita Grasslands	Lab	47
		0.25	Melfort sic	Lab	46
	0.14		Wilson c	Field	48
	0.06		Crocett sl	Field	48
Benzonitriles					
Bromoxynil	0.203		Stondon Masseyu cl	Field	49
	0.274		Beck row s	Field	49
Chlorthiamid	0.025—0.1		sl	Field	50
Dichlobenil	>0.012 and 0.002		sl	Field	11
		>0.012 and 0.0021	sl	Field	11
	0.05—0.1		sl	Field	50
Bypyridyls					
Paraquat	<0.002	<0.002	sl	Field	51
Carbamates					
Asulam		0.046	cl	Lab	52
Diphenyl Ethers					
Fluorodifen		0.015—0.17	Lufkin sl	Field	53
		0.013	Miller c	Field	53
Nitrofen	0.154		Plainfield s	Field	54
Dinitroanilines					
Benefin		0.0147	Plano si l	Field	55
Butralin		0.024	Bosket sl	Lab	56
		0.013	Sharkey c	Lab	56
Dinitramine		0.02	Bosket sl	Lab	56
		0.015	Sharkey c	Lab	56
		0.0084	cl Aridic Arguistoll	Field	57

Table 4 (continued)
SELECTED RATE (k_s) VALUES (PER DAY) FOR DISSIPATION OF
HERBICIDES FROM SOILS, PRIMARILY FIELD

Pesticide	Dissipation rate (k_s)		Soil type	Experimental	Ref.
	Surface	Rhizosphere			
Dinitroanilines					
Ethalfluralin		0.023	Regina heavy c	Field	58
		0.022	White City sl	Field	58
Fluchloralin		0.024	Bosket sl	Lab	56
		0.0095	Sharkey c	Lab	56
		0.0068—0.013	Taloka si l	Field	6
		0.0068	Crowley si l	Field	6
	0.099		Sharkey c	Field	35
	0.126		Bosket sl	Field	35
		0.023	Plano si l	Field	55
Isopropalin		0.011	Plano si l	Field	55
Nitralin		0.007	Dundel si cl	Lab	59
		0.008	Ochley si l	Field	39
Oryzalin		0.02	Plano si l	Field	55
Pendimethalin	0.006		Crowley si l	Field	60
		0.002—0.01	cl	Lab	61
		0.011	sl	Lab	61
		0.027	c	Lab	61
	0.054		Sharkey c	Field	35
	0.082		Bosket sl	Field	35
	0.17		Taloka si l	Field	35
Profluralin		0.012	Bosket sl	Lab	56
		0.0056	Sharkey c	Lab	56
		0.020	Plano si l	Field	55
Trifluralin	0.061		Elkton si l	Field	62
		0.004	Several loams	Field	63
		0.0093	Ochley si l	Field	39
		0.0082—0.062	Hidalgol sl	Field	42
		0.013	Regina heavy c	Field	64
		0.0091	White City sl	Field	65
		0.0074	Regina heavy c	Field	65
		0.02	Regina heavy c	Field	58
		0.019	White City sl	Field	58
		0.0034	sl	Field	44
	0.009—0.016		Bosket sl	Field	66
	0.035		Cecil	Field	67
Phenoxyalkanoic Acids					
2,4-D	0.062		Galestown sl	Chambers	5
		0.059	Heavy c	Lab	68
MCPA		<0.01	cl, Heavy c, sl	Lab	69
		0.053	cl, Heavy c, sl	Lab	52
Mecoprop		0.087	cl, Heavy c, sl	Lab	69
Silvex®	0.016—0.034		Galestown sl	Chambers	5
2,4,5-T	0.042—0.050		Chester l, Chandler fsl	Field	70
		0.034	Heavy c	Lab	68

Table 4 (continued)
SELECTED RATE (k_s) VALUES (PER DAY) FOR DISSIPATION OF HERBICIDES FROM SOILS, PRIMARILY FIELD

Pesticide	Dissipation rate (k_s) Surface	Rhizosphere	Soil type	Experimental	Ref.
Thiocarbamates					
Butylate		0.0056	Tripp fsl	Lab	71
Cycloate		0.032	Tripp fsl	Lab	71
Diallate		0.013—0.0233	Weyburn	Lab	72
		0.013—0.016	Regina heavy c	Field	72
EPTC		0.039—0.077	Kennebes si l	Field	73
		0.046—0.116	Hastings si l	Field	73
		0.031	Tripp fsl	Lab	71
		>0.026	Regina heavy c	Field	45
		>0.025	White City sl	Field	45
		>0.02	Melfort cl	Field	45
Triallate		0.0056	Regina heavy c	Lab	72
		0.0058—0.011	Weyburn l	Lab	72
		0.0096	White City sl	Field	65
		0.0071	Regina heavy c	Field	65
Vernolate		0.033	Tripp fsl	Lab	71
Triazines					
Atrazine	0.021—0.025		Maury si l	Field	74
		0.0123	Brookston cl	Field	75
	0.013—0.019		l	Field	43
	0.019—0.02		sl	Field	43
		0.0173	sl	Field	44
		0.024	Collington fsl	Field	41
	0.007		Holdrege 1—conventional-till	Field	76
	0.006		Holdrege 1—no till	Field	76
	0.018—0.030		Decatur si l pH depend.	Field	77
	0.031—0.042		Hartsells fsl pH depend.	Field	77
	0.060—0.085		McLaurin sl pH depend.	Field	77
	0.030		Bloomfield fs	Field	78
	0.02		Ochley si l	Field	78
	0.012		Brookston cl	Field	75
	0.027—0.046		Chalmers si c, trace si l	Field	79
Cyanazine	0.041		Bloomfield fs	Field	78
	0.02		Ochley si l	Field	78
	0.016		sl	Field	80
Metamitron		0.031	sl	Field	44
Metribuzin		0.028—0.032	Bosket sl	Pots	81
		0.024—0.041	Sharkey c	Pots	81
		0.032	Forestdale cl	Pots	81
		0.036	Pearson si l	Pots	81
	0.013—0.017		Hillsdale sl	Field	82
	0.055—0.131		Cecil sl	Field	83
	>0.017		Regina heavy c	Field	45
	>0.0016		Melfort cl	Field	45
	>0.023		White City sl	Field	45

Table 4 (continued)
**SELECTED RATE (k_s) VALUES (PER DAY) FOR DISSIPATION OF
HERBICIDES FROM SOILS, PRIMARILY FIELD**

Pesticide	Surface	Rhizosphere	Soil type	Experimental	Ref.
	Dissipation rate (k_s)				
			Triazines		
Metribuzin	0.027		sl	Field	44
Procyazine	0.055		sl	Field	80
Prometryn	0.0085—0.014		sl	Field	84
Simazine	0.010—0.015		sl	Field	84
		0.011	sl	Field	44
	0.0064—0.06		World soils	Field	85
			Uracils		
Lenacil	0.011			Field	87
Terbacil	0.0033—0.0046		Fox sl	Field	86
			Ureas		
Chlorbromuron	0.012		New Brunswick	Field	88
Diuron		0.0083—0.011	Decatur cl	Lab	89
		0.011	Norfolk ls	Lab	89
		0.0022	Fox ls	Field	90
Linuron		0.0059	Norfolk sl	Field	44
		0.0055	l	Lab	43
	0.014—0.012		l	Field	43
	0.01		si l	Field	43
		0.013—0.053	l	Lab	91
Monuron		0.003	Ramona sl	Field	92
Neburon		0.004	Ramona sl	Field	92
Tebuthiuron	0.0016		Rillito and Laveen	Field	93
			Miscellaneous		
Chlorsulfuron	0.024		sl	Field	94
DCPA	0.059		sl	Field	44
Fluridone	0.0041		Lufkin fsl	Field	95
		0.0034	Lufkin fsl	Field	95
Glyphosate		0.028	Lintonia sl	Lab	96
		0.02	Drummer si cl	Lab	96
		0.0006	Norfolk sl	Lab	96
		0.022	Raye si l	Lab	96
Oxadiazon	0.006—0.11		Crowley si l	Field	97
Picloram	0.003—0.007		sl	Field	98
	0.026		Chandler fsl	Field	70
	0.027		Chester l, Pannin cl	Field	70
		0.086	Three loams	Field	99
Sethoxydim		0.05	cl	Lab	100
		0.059	sl	Lab	100
		0.022	Heavy c	Lab	100

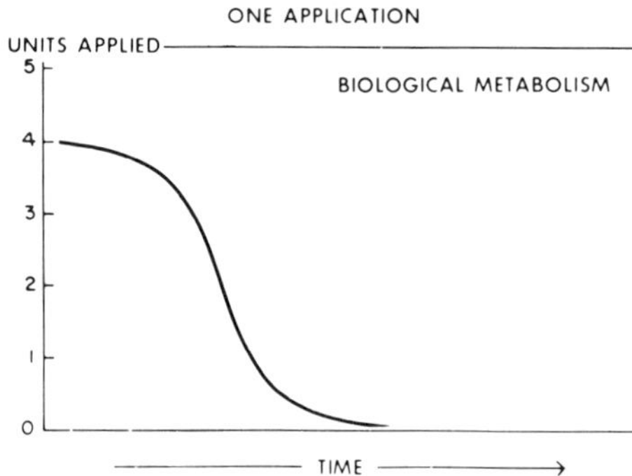

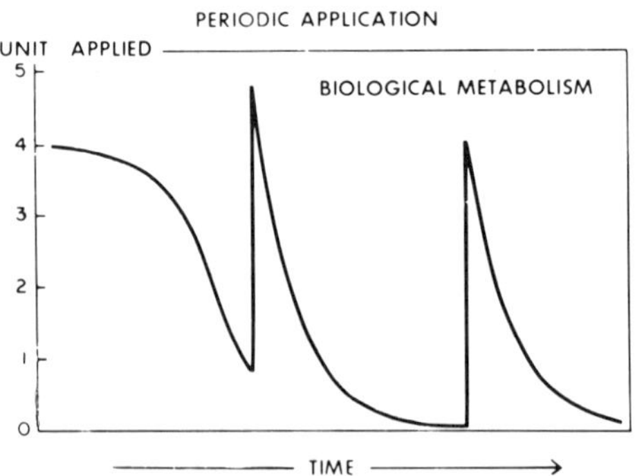

FIGURE 4.   Pesticide losses (dissipation) through biological degradation from one or more applications. (From Kearney, P. C., Nash, R. G., and Isensee, A. R., in *Chemical Fallout: Current Research on Persistent Pesticides,* Miller, M. W. and Berg, G. G., Eds., Charles C. Thomas, Springfield, Illinois, 1969. With permission.)

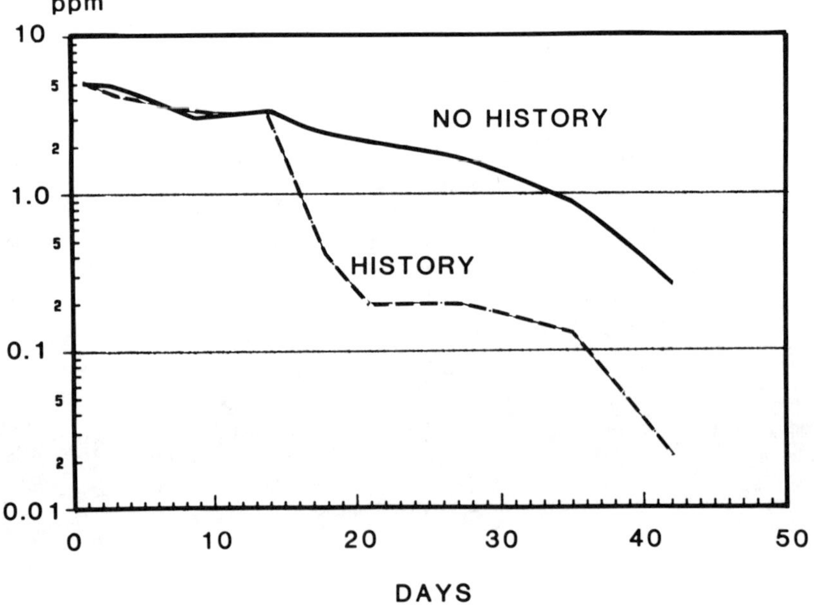

FIGURE 5.   Rate of EPTC degradation of 6.7 kg/ha on a Kennebec silt loam soil with a prior history of EPTC application vs. a soil with no prior history. (In this case the lag period for the history soil probably was caused by the soil being dried prior to the EPTC application.) (From Obrigawitch, T., Martin, A. R., and Roeth, F. W., *Weed Sci.,* 30, 175, 1982. With permission.)

mechanisms, such as a microbial population buildup or enzymatic adaptation, we would expect the dissipation curve to be like that depicted in Figure 4.[30] Initially, dissipation would be slow, but as an adaptive microbial population builds up or an enzyme adapts, the dissipation rate would increase such that the situation would be as shown in Figure 5.[31] In Figure 5, the adaptive microbial population was already present, but was in a resting state because the soil had been dried.

When dissipation is primarily microbial, soil temperature and moisture influence dissipation considerably. At low soil temperatures and moisture content, degradation essentially ceases. Likewise, at high soil temperatures (soil surface) microbial degradation presumably ceases. At high soil moisture content (saturation) and a favorable soil temperature, however, degradation may cease or continue by a different pathway. Therefore, dissipation from microbial degradation may be more like a "stop and go" process in many agricultural soils rather than continuous. This would be true especially for herbicides on the soil surface. Nevertheless, the overall dissipation rate effect usually is pseudo first order.

Some of the dissipation processes listed in Figure 2A (most physical and chemical and photochemical transformations and degradations) do occur as a first-order rate process under ideal conditions. Other processes (volatilization and some microbial degradations) also give the appearance of first-order rate losses.[3,32] When combined, as in field soil, the pesticide loss from the several dissipation processes nearly always follows pseudo first-order kinetics when soil and temperature conditions are favorable.

Use of pseudo first-order kinetics greatly simplified the development of equations for describing or estimating pesticide loss from soil and allows the use of the kinetic concept of half-life (referred to as half-concentration time, $C_{1/2}$ in this chapter).

The pseudo first-order equation is given by the following expression:

$$C_d = C_o\, e^{-k_s d} \tag{1}$$

where C_d = herbicide concentration on d = day, C_o = initial herbicide concentration or concentration on d = 0, e = 2.718, and k_s = dissipation rate constant. Although use of generalized Equation 1 to estimate dissipation from soil is useful, the specifically derived equation is theoretically valid only for the particular data set from which it was derived. For example, Equation 1 calibrated to the observed dissipation of a herbicide incorporated into a cold soil would very much underestimate herbicide dissipation from a warm soil.

In hopes of overcoming the shortcomings (e.g., temperature effects) of Equation 1, the weather variables have been incorporated in dissipation equations.[14,33] A convenient way to do this is to cumulate the weather variables on a daily basis (i.e., daily mean air temperature [AT], wind run [R{maximum or minimum may be useful under certain circumstances also}], and relative humidity [RH] or daily rainfall [R]):

$$C_d = C_o\, e^{-k_{s1}\ AT} \tag{2}$$

Equation 2 will be referred to as an integral-variable equation. Further, Equation 2 can be expanded to include more than one environmental factor through stepwise regression:

$$C_d = C_o\, e^{-k_{s1}\ AT\ +\ k_{s2}\ R} \tag{3}$$

Fortunately, the availability of automatic environmental recording devices and computer statistical analyses has greatly reduced the work of factoring weather data into the equations. Unfortunately, very few data sets in existence have the necessary environmental variables available for use in equation 3 which limits its use. I have used cumulative mean daily air, surface soil, and rhizosphere soil temperatures; cumulative mean daily relative humidity and wind run; cumulative pan evaporation (centimeters of water evaporated from an open pan); rainfall; and number of rain incidences as weather variables in Equation 3.[34] Other parameters could also be included. The importance of each weather variable was determined with a computer by stepwise regression statistical analysis, which usually identified the one or two variables that explained much of the variation.

Occasionally, logarithmic concentration by logarithmic time equations describes pesticide dissipation from soil.[14] The logarithmic concentration and logarithmic time equation takes the form:

$$\ln C_d = \ln C_o - kd$$

or

$$C_d - C_o d^{-k} \tag{4}$$

Further observations have indicated that a single rate is often inadequate to describe dissipation from soil surfaces.[8,26] Therefore, a two-compartment exponential model has been proposed.[8]

$$C_d = C_o\, e^{-(k_s + k_r)d} + C_o \frac{k_r}{(k_s + k_r - k_d)} [e^{-k_d d} - e^{-(k_s + k_r)d}] \tag{5}$$

where K_r = retained residue and k_d = degradation rates. Testing of the two-compartment model on a few data sets indicates it has considerable validity. Further, Hill and Schaalje[8] included °day in the model, which helps to reduce some of the temperature fluctuation errors.

Leistra has outlined the pseudo first-order dissipation rate of herbicides that transform stepwise to other products during dissipation.[27]

$$
\begin{array}{ccccccc}
 & \underline{\quad} k_{12} \underline{\quad} & C_2 & \underline{\quad} k_{23} \underline{\quad} & \\
C_1 & & | & & C_3 \\
| \; k_{14} & & | \; k_{24} & & | \; k_{34}
\end{array}
$$

$$\text{collection of products—} C_4$$

C_1 is the parent herbicide and C_2 and C_3 are transformation products (e.g., oxidation products), whereas C_4 may be the end products from degradation. In this scheme it was assumed that first-order or pseudo first-order kinetics applied; therefore, the following set of differential equations was obtained:

$$dC_1/dt = -k_{12}C_1 - k_{14}C_1 \tag{6}$$

$$dC_2/dt = k_{12}C_1 - k_{23}C_2 - k_{24}C_2 \tag{7}$$

$$dC_3/dt = k_{23}C_2 - k_{34}C_3 \tag{8}$$

$$dC_4/dt = k_{14}C_1 - k_{24}C_2 + k_{34}C_3 \tag{9}$$

The set of equations was solved numerically. For example, the dissipation of C_1 was found from the sum of the rate coefficients ($k_{12} + k_{14}$) and C_2 from ($k_{23} + k_{24}$). When any one k value was determined, it was possible to determine the other paired value.

2. Dissipation Pathways and Rate Constants (k)

Soils are not static, especially during the summer. Pesticides volatilize with little of the vaporous pesticide returning to the soil. Pesticides hydrolyze irreversibly. Various other transformations and degradation processes occur, which all reduce C_d. These processes usually go in one direction and usually are relatively slow compared with equilibration, i.e., the partitioning and dissociation. Therefore, a computational approach to estimating C_d, as opposed to the empirical/experimental approach, can also be done:[36]

$$C_d = C_o\, e^{-(k_a + k_p + k_h + k_b + k_o + k_r}$$

$$+ k_l + k_f + k_c + k_{bcf} - k_d - k_e)d \tag{10}$$

where k_a, k_p, k_h, k_b, k_o, k_r, k_l, k_f, k_c, k_{bcf}, k_d, k_e = volatilization, photolysis, hydrolysis, biodegradation, oxidation, reduction, leaching, diffusion, complexation, bioconcentration factor (bioaccumulation by soil fauna and flora), desorption, and bioelimination rates, respectively.

Volatilization (k_a) is considered a one-way process, although condensation back onto and into the soil from herbicide vapors has been demonstrated over very short (5 cm) distances.[37] Bioelimination (k_e) (i.e., from root exudation of foliar-applied and stem-translocated pesticide and physically introduced to soil by mobile soil fauna) seemingly could enrich the soil pesticide pool, but probably is not a measurable factor and can be omitted from the equation. Likewise, desorption (k_d) from "bound" residues could enrich the available pool, but most likely would be rate controlling if its occurrence was measurable. Therefore, for all practical purposes, desorption can be omitted. Photolysis (k_p) is of little importance, except for certain photosensitive surface-applied pesticides. Oxidation (k_o) and reduction (k_r) depict a transitory, difficulty-predictable state, which overall may or may not enhance dissipation. Leaching (k_l) and diffusion (k_f) are pesticide transport from one component (surface, rhizosphere, subsoil) to another. In Equation 6, leaching (k_l) and diffusion (k_f) are considered to be decreases in pesticide amount in the soil component under consideration (i.e., soil surface or rhizosphere). But, when gaseous diffusion is not to the atmosphere, then both leaching and diffusion would carry a $(-)$ sign in the equation for the component (soil surface, rhizosphere, or subsoil) into which the pesticide is being transported. Complexation (k_c) or formation of "bound" residues, which in most cases for all practical purposes removes the pesticide from the available soil pool, is a dissipation process.

Dissipation mechanisms are beyond the scope of this chapter and have been discussed elsewhere.[3,36] It will suffice here to mention briefly the several dissipation pathways, except for volatilization, because more progress has been made in estimating pesticide volatilization from physicochemical and soil or environmental properties. Many soil pesticide losses are not true kinetics, because of the complexity of the conditions in the environment.

An extensive literature survey[3] and regression analysis of dissipation data, however, more often than not indicate that a lumped sum of all these various dissipation (k) pathways gives the appearance of a pseudo first-order equation with a high degree of statistical significance.

3. Effect of Herbicide
Other than for very cold or dry soils, the single most important factor affecting herbicide dissipation is the herbicide itself (Figure 6). Some herbicides dissipate from the soil rapidly and others remain for a long time. Benefin, for example, had a half-concentration time ($C_{1/2}$) of 2 days, while fluchloralin had a $C_{1/2}$ of 41 days, both on the same soil.[38] Further, various dissipation rates can occur with the same herbicide on the same soil under the same conditions.[39]

4. Dissipation by Class
Table 4[5,6,11,32,35,39-100] lists the persistence ($C_{1/2}$ = time in days required for 50% of the herbicide to dissipate) and rate constant (k_s) from pseudo first-order kinetics of several herbicides in or on various soils in the northern hemisphere. Although it is not demonstrated in Table 4, k_s values obtained from laboratory data tend to be smaller than those obtained from field data.

Therefore, persistence in the field, except for cold or frozen soils, usually is shorter than that demonstrated from laboratory experiments.

D. Soil and Climatic Parameters
Table 1 lists some of the soil and climatic parameters and variables.

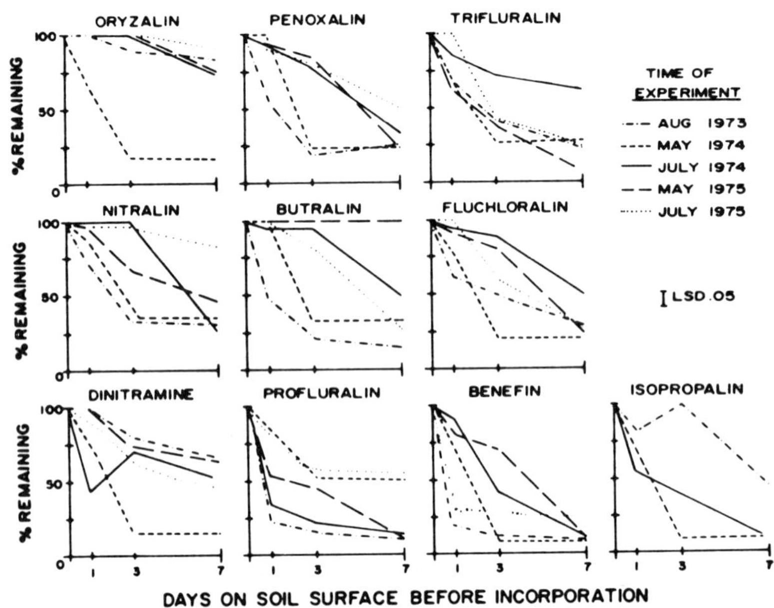

FIGURE 6.   Amount of various dinitroaniline type herbicides remaining on a Taloka s
1 soil with 10% organic matter. Average maximum soil surface temperatures (°C): Aug.
1973 — 48, May 1974 — 35, July 1974 — 46, May 1975 — 40, and July 1975 — 47.
Total precipitations (cm) during the experiments were: Aug. 1973 — 3.45, May 1974
— 11.7, and trace or none for the remainder. (From Kennedy, J. M. and Talbert, R. E.,
Weed Sci., 25, 373, 1977. With permission.)

1. Soil Type

Numerous examples exist demonstrating the effect of soil type on the persistence of
herbicides in or on soil (Table 4, Table 5,[73,101-104] Table 6,[41] Table 7,[6,26,40,43,44,57,85,94,101-103,105-107]). No general statement can be made about the effect of soil type on herbicide
dissipation, except perhaps that very high organic matter soils tend to decrease the dissipation.
Although soil type is an important factor governing herbicide persistence, other factors, such
as soil temperature and soil moisture (and sometimes soil pH), usually have a greater effect
when varied over a wide range.

The data of Nash[36] indicated that of the soil and climatic variables affecting dissipation,
temperature (cumulative daily for soils or air), soil moisture (cumulative daily), and rain
(cumulative daily for rain amount or incidences) are the most highly correlated with pesticide
dissipation. Rain incidences perhaps are more important for pesticides remaining on the soil
surface, and cumulative rain perhaps was the most important variable in one data set.

2. Soil Temperature

Soil temperature or time of year affects herbicide dissipation (Tables 5 to 7, Figure 6).
Generally, within field soil temperature ranges, the higher the temperature, the more rapid
the dissipation (Tables 5, 6). For all practical purposes, dissipation is presumed to cease at
temperatures of freezing and below.

Data for picloram dissipation from soils in several locations in the U.S. and Canada show
the effect of temperature also.[26] Figure 7 is a plot of latitude vs. $k_{1/2}$ (the half-order equation
was used to describe the dissipation of picloram rather than the first-order equation, because
only two data points were collected). In the northern hemisphere, the $k_{1/2}$ values decreased
from north to south, which indicated a more rapid dissipation of picloram from north to
south. For example, the estimated $C_{1/2}$ for picloram at 25 and 40° latitude is 29 and 113
days, respectively.

Table 5
EFFECT OF SOIL TYPE, ORGANIC MATTER CONTENT, MOISTURE CONTENT, TEMPERATURE, AND pH ON DISSIPATION (k_s VALUES — PER DAY) OF HERBICIDES IN RHIZOSPHERE SOIL

| | Soil | | | 5°C | | 10°C | | 14—15°C | | 20°C | | 25°C | | 28—30°C | | 35°C | | |
| | | | | | | | | | | | | | | | | | | |
Pesticide	Type	pH	OM[a]	SM[b] (%)	k_s	SM (%)	k_s	SM (%)	k_s	SM (%)	k^s	SM (%)	k_s	SM (%)	k_s	SM (%)	k_s	Ref.
Asulam	Regina c	7.7	7.2			20	0.0141	20	0.0261	20	0.0440	20	0.0590	20	0.0582	20	0.0774	101
	Regina c	7.7	7.2			26	0.0290	26	0.0397	26	0.0564	26	0.0723	26	0.0596	26	0.0792	101
	Regina c	7.7	7.2			34	0.0358	34	0.0504	34	0.0981	34	0.0957	34	0.1052	34	0.0814	101
	Regina c	7.7	7.2			40	0.0386	40	0.0525	40	0.1127	40	0.0970	40	0.1174	40	0.0868	101
EPTC	Tripp vfsl	8.0	1.0									0.5	0.015**[c]					73
	Tripp vfsl											1.6	0.0144**					73
	Tripp vfsl											3.0	0.055**					73
	Tripp vfsl											13.0	0.66**					73
	Tripp vfsl											16.0	0.72**					73
	Kennebec si l[d]	6.6	2.4	100	0.0201**			10	0.0604**			10.0	0.063**					73
	Kennebec si l[e]	6.6	2.4	100	0.0763**			10	0.312			10.0	0.443					73
Linuron	Gravel Pits sl	7.0	2.0	4.0	0.0065	4.0	0.0083	4.0	0.0104	4.0	0.0122	4.0	0.154	4.0	0.0293			102
	Gravel Pits sl	7.0	2.0									8.0	0.0165					102
	Gravel Pits sl	7.0	2.0	12.0	0.0084	12.0	0.0110	12.0	0.0131	12.0	0.0162	12.0	0.0187	12.0	0.0193			102
	Gravel Pits sl	7.0	2.0									16.0	0.0204					102
Prometryn	Gravel Pits sl	7.0	2.0									11.2	0.0176					103
	Gravel Pits sl	7.0	2.0					11.5	0.0062			11.4	0.0191					103
	Gravel Pits sl	7.0	2.0									13.2	0.0188					103
	Gravel Pits sl	7.0	2.0									4.8	0.0012					103
	Gravel Pits sl	7.0	2.0					6.6	0.0014			6.0	0.0029					103
	Gravel Pits sl	7.0	2.0									7.9	0.0084					103
	Gravel Pits sl	7.0	2.0									9.7	0.0152					103
	Gravel Pits sl	7.0	2.0									10.7	0.0167					103
	Gravel Pits sl	7.0	2.0									11.2	0.0234					103
	Gravel Pits sl	7.0	2.0					11.5	0.0062			11.4	0.0276					103
	Gravel Pits sl	7.0	2.0									13.2	0.0229					103
Pronamide	Gravel Pits sl	6.1	1.8					11.7	0.0091			4.1	0.0116					104
	Little Cherry sl	6.9	1.4					12.0	0.0099			4.0	0.0096					104
	Soakwaters cl	6.1	1.4					17.8	0.0110			6.4	0.0124					104
	Gallas Ley sc	6.9	2.9					16.7	0.0062			6.2	0.0074					104
	Water Meadows cl	7.2	5.3					29.5	0.0082			14.8	0.0142					104
	Gravel Pits sl	6.1	1.8									11.9	0.0217					104
	Little Cherry sl	6.9	1.4									12.1	0.0231					104
	Soakwaters cl	6.1	1.4									16.3	0.0301					104

Table 5 (continued)
EFFECT OF SOIL TYPE, ORGANIC MATTER CONTENT, MOISTURE CONTENT, TEMPERATURE, AND pH ON DISSIPATION (k_s VALUES — PER DAY) OF HERBICIDES IN RHIZOSPHERE SOIL

| | Soil | | | Temperature | | | | | | | | | | | | |
| | | | | 5°C | | 10°C | | 14—15°C | | 20°C | | 25°C | | 28—30°C | | 35°C | | |
Pesticide	Type	pH	OM[a]	SM[b] (%)	k_s	SM (%)	k_s	SM (%)	k_s	SM (%)	k^s	SM (%)	k_s	SM (%)	k_s	SM (%)	k_s	Ref.
Pronamide	Gallas Ley sl	6.9	2.9									16.7	0.0165					104
	Water Meadows cl	7.2	5.3									29.1	0.0198					104
	Little Cherry sl	6.1	2.0					7.5	0.0062					3.5	0.0077			104
	Little Cherry sl	6.1	2.0											7.5	0.0110			104
	Little Cherry sl	6.1	2.0											10.0	0.0127			104
Simazine	Gravel Pits sl	7.0	2.0	4.0	0.0033	4.0	0.0050	4.0	0.0075	4.0	0.0116	4.0	0.0173	4.0	0.0239			102
	Gravel Pits sl	7.0	2.0									8.0	0.0248					102
	Gravel Pits sl	7.0	2.0	12.0	0.0056	12.0	0.0087	12.0	0.0131	12.0	0.0204	12.0	0.0267	12.0	0.0433			102
	Gravel Pits sl	7.0	2.0									16.0	0.0301					102
	Gravel Pits sl	7.0	2.0									4.8	0.0082					103
	Gravel Pits sl	7.0	2.0					6.6	0.0030			6.0	0.0091					103
	Gravel Pits sl	7.0	2.0									7.9	0.0109					103
	Gravel Pits sl	7.0	2.0									9.7	0.0131					103
	Gravel Pits sl	7.0	2.0									10.7	0.0149					103

[a] Organic matter
[b] Soil moisture.
[c] 99% confidence level.
[d] No previous EPTC history.
[e] Previous EPTC history.

Table 6
DISSIPATION RATE (k_s) (PER DAY) OF AMIDE HERBICIDES AS A FUNCTION OF SOIL TYPE, TEMPERATURE, AND MOISTURE

Conditions		Dissipation rate[a]					
		Alachlor		Metolachlor		Propachlor	
Temperature (°C)	Soil moisture field cap. (%)	sl[b]	cl[b]	sl	cl	sl	cl
Temperature Experiment							
10	50	0.018a	0.046a	0.007a	0.025a	0.042a	0.049a
20	50	0.028a	0.053a	0.014ab	0.025a	0.210b	0.131b
30	50	0.045b	0.116b	0.016b	0.053b	0.365c	0.693c
Moisture Experiment							
20	20	0.016c	0.028c	0.007c	0.018c	0.030d	0.032d
20	50	0.028d	0.053d	0.014cd	0.025cd	0.231e	0.131c
20	80	0.037d	0.062d	0.021d	0.044d	0.231e	0.169e

[a]   Values in both columns for one herbicide followed by the same letter were not significantly different at the 95% level by Student t-test for comparison of two linear regressions.

[b]   sl = Aridic Argiustoll soil with 2.5% organic matter and pH of 8.0.
cl = Ustollic Haplargid soil with 1.1% organic matter and pH of 7.8.

Adapted from Zimdahl, R. L. and Clark, S. K., *Weed Sci.*, 30, 545, 1982.

3. Soil Moisture

Soil moisture affects the dissipation of most pesticides; dissipation is more rapid from moist than dry soils (Tables 5, 6).[26,28] The extensive work of Walker and Zimdahl and coworkers shows the influence of moisture on the dissipation of herbicides.[41,43,44,57,85,94,101-103,106] Usually, the drier the soil the smaller the k_s value, indicating a longer $C_{1/2}$ time, and the wetter the soil, at least to near field capacity (33 kPa soil moisture tension), the bigger the k_s and the shorter the $C_{1/2}$ time. The $C_{1/2}$ of trifluralin was near 100 days on a dry soil, but only 7 days on a moist soil.[108] The k_s value for propham was 0.025 on dry soil and 0.279 on wet soil, which gave $C_{1/2}$ of 27 and 2.5 days, respectively.[109]

4. Soil pH

Soil pH affects the dissipation of several herbicides, although no general conclusion can be made (Tables 5, 8). Some herbicides are less stable at low pH values and some at high pH values, depending on their functional groups.

5. Soil Organic Matter

Soil organic matter can affect the dissipation of a herbicide in two ways. First, increased adsorption of the herbicide onto the organic matter may decrease its rate of dissipation.[110] This may be especially true where volatilization of the herbicide is a major pathway of dissipation. If, however, microbial degradation is a major dissipation pathway, then enhanced microbial activity associated with higher soil organic matter content may increase the rate of dissipation. Soil organic matter effects on herbicidal dissipation probably cannot be separated from soil type (Tables 5,6).

E. Soil Surface

Dissipation of most herbicides is more rapid from surface soil than from rhizosphere

Table 7
CALCULATED DISSIPATION ACTIVATION ENERGIES (ΔE) AND SOIL-MOISTURE INFLUENCE (B), WHICH DESCRIBE HERBICIDE DISSIPATION FROM SOIL

Herbicide	Conditions	ΔE Equation		Soil moisture equation		Ref.
		SM[a] SMT[b] %	kJ mol^{-1}	°C	B[c]	
Alachlor	sl Rhizosphere, 10—30°C	8.4[a]	28.9 ± 0.6	20	0.60 ± 0.01	40
	cl Rhizosphere, 10—30°C	11.5[a]	32.4 ± 13.9	20	0.61 ± 0.09	40
Ametryn	13.2—31.2°C		74.9			26
	13.2—31.2°C		75.9			26
Amitrole	8—35°C		26.1			26
	35—100°C		12.8			26
Asulam	Regina c rhizosphere, 10—35°C	20[a]	46.8 ± 7.8**[d]			101
	Regina c rhizosphere, 10—35°C	26[a]	27.1 ± 5.8**			101
	Regina c rhizosphere, 10—30°C	34[a]	41.7 ± 9.3**			101
	Regina c rhizosphere, 10—30°C	40[a]	42.3 ± 10.8			101
	Regina c rhizosphere, 20—40% SM			10	1.41 ± 0.37*[d]	101
	Regina c rhizosphere, 20—40%SM			15	1.02 ± 0.19**	101
	Regina c rhizosphere, 10—40°C			20	1.45 ± 0.17**	101
	Regina c rhizosphere, 10—40°C			25	0.77 ± 0.11**	101
	Regina c rhizosphere, 10—40°C			30	1.14 ± 0.29*	101
	Regina c rhizosphere, 10—40°C			40	0.15 ± 0.04[ns,d]	101
Atrazine	3 Soils, rhizosphere, 5—25°C	70[b]	49.0 ± 3.2			43
	3 Soils, rhizosphere, 70% SMT			25	0.63 ± 0.15	43
	sl Rhizosphere	60[b]	69.9 ± 4.5	20	0.16 ± 0.10	44
Bromacil	13.2—31.2°C		89.1			26
Chlorsulfuron	s1,10—30°C	12[a]	58.9			94
	sl, 6—12% SM			20—30	1.12	94
Chlorthal-dimethyl			92.0 = 5.5		0.99 ± 0.13	44
Chloroxuron	13.2—31.2°C		38.9			26
Dinitramine	cl Rhizosphere, 10—30°C	17[a]	96.6 ± 35.9			57

	Wickham sl rhizosphere, 10—30°C	9.6[a]	88.7 ± 19.9			57
	cl Rhizosphere, 2.2—22% SM			30	1.06 ± 0.17	57
	Wickham sl rhizosphere, 0.5—12% SM			30	0.99 ± 0.11	57
Diuron	6.5—31.2°C		44.0			26
Fenuron	13.2—31.2°C		6.2			26
Fluchloralin	Sharkey si c rhizosphere, 4—25°C	34[a]	33.6			6
	Crowley si l rhizosphere, 4—25°C	27[a]	40.7			6
Isopropalin	Aerobic weld l, 15—30°C	20[a]	65.0			105
	Anaerobic weld l, 15—30°C	20[a]	66.0			105
Linuron	2 Soils, 5—25°C	70[b]	36.7 ± 1.1	25	0.84 ± 0.04	43
	sl Rhizosphere	60[b]	28.8 ± 3.8	20	0.47 ± 0.03	44
	Gravel Pits sl rhizosphere, 5—30°C	4[a]	38.5 ± 6.1**			102
	Gravel Pits sl rhizosphere, 5—30°C	12[a]	24.6 ± 2.0**			102
	Gravel Pits sl rhizosphere, 4—12% SM			5	0.24 ± 0.01	102
	Gravel Pits sl rhizosphere, 4—12% SM			10	0.24 ± 0.17	102
	Gravel Pits sl rhizosphere, 4—12% SM			15	0.21 ± 0.01	102
	Gravel Pits sl rhizosphere, 4—12% SM			20	0.25 ± 0.01	102
	21.5—26°C		38.9			26
	Gravel Pits sl rhizosphere, 4—16% SM			25	0.20 ± 0.04**	102
	Gravel Pits sl rhizosphere, 4—16% SM			30	−0.38 ± 0.01[e]	102
Metamitron	2 Soils, 5—25°C	60[b]	46.6 ± 5.1	20	1.00 ± 0.03	44
Metolachlor	3 Soil rhizosphere, 5—25°C	70[b]	48.5 ± 3.8	25	0.95 ± 0.08	43
	sl Rhizosphere, 10—30°C	8.5[a]	30.4 ± 10.5	20	0.79 ± 0.03	40
	cl Rhizosphere, 10—30°C	11.5[a]	26.0 ± 15.7	20	0.59 ± 0.21	40
Metribuzin	sl Rhizosphere	60[b]	65.0 ± 4.4	20	0.71 ± 0.09	44
Monolinuron	21.5—26°C		132.6			26
Monuron	13.2—31.2°C		31.9			26
Oryzalin	Aerobic weld l, 15—30°C	20[a]	72.0[b]			105
Profluralin	Crowley si l rhizosphere, 4—25°C	27[a]	35.6			6
	Sharkey si c rhizosphere	34[a]	35.4			6
Prometryn	sl			25	2.76 ± 0.28	103
	Gravel Pits sl rhizosphere, 15—25°C	11.5[a]	93.4 ± 13.1			103
		6—6.6[a]	52.1 ± 5.7			103
	6.6—11.5% SM			15	2.63 ± 0.06**	103
	4.8—13.2% SM			25	2.93 ± 0.26**	103

Table 7 (continued)

CALCULATED DISSIPATION ACTIVATION ENERGIES (ΔE) AND SOIL-MOISTURE INFLUENCE (B), WHICH DESCRIBE HERBICIDE DISSIPATION FROM SOIL

		ΔE Equation		Soil moisture equation		
Herbicide	Conditions	SM[a] SMT[b] %	kJ mol^{-1}	°C	B[c]	Ref.
Pronamide	Gravel Pits sl rhizosphere, 15—25°C	11.8[a]	50.6 ± 8.1			106
	Little Cherry sl rhizosphere, 15—25°C	12.0[a]	60.6 ± 0.8			106
	Little Cherry sl rhizosphere, 15—30°C	7.5[a]	27.5 ± 4.5			106
	Soakwaters cl, 15—25°C	16.3—17.8[a]	79.2 ± 0.7			106
	Gallas Ley sc, 15—25°C	16.7[a]	68.9 ± 1.3			106
	Water Meadows cl, 15—25°C	29.5[a]	62.9 ± 0.9			106
	Little Cherry sl, 7.5—12% SM			15	1.0	106
	Little Cherry sl, 3.5—10% SM			30	0.47	106
	Gravel Pits sl, 4.1—11.9% SM			25	2.6	106
	Little Cherry sl, 4.0—12.1% SM			25	0.79	106
	Soakwaters cl, 6.4—16.3% SM			25	0.95	106
	Gallas Ley sc, 6.2—16.7% SM			25	0.80	106
	Water Meadows cl, 14.8—29.1% SM			25	0.49	106
	sl Rhizosphere	60[b]	74.6 ± 5.5	20	0.68 ± 0.15	44
Propachlor	sl Rhizosphere, 10—30°C	8.5[a]	77.9 ± 20.5	20	1.50 ± 0.53	40
	cl Rhizosphere, 10—30°C	11.5[a]	94.5 ± 15.6	20	1.24 ± 0.25	40
Simazine	sl Rhizosphere, 10—30°C	60[b]	69.5 ± 3.2	20	0.55 ± 0.09	44
	16 Soil surfaces, 10—30°C	20—90[b]	48.2 ± 12.1	20	0.46 ± 0.41	85
	8.5—25°C		61.7			107
	Gravel Pits sl,5—30°C	4[a]	56.3 ± 0.8**			102
	Gravel Pits sl, 5—30°C	12[a]	56.2 ± 1.6**			102
	Gravel Pits sl, 15—25°C	6—6.6[a]	79.2 ± 2.5			102
	Gravel Pits sl, 4—12% SM			5	0.48 ± 0.03	102
	Gravel Pits sl, 4—12% SM			10	0.50 ± 0.02	102
	Gravel Pits sl, 4—12% SM			15	0.57 ± 0.02	102
	Gravel Pits sl, 4—12% SM			20	0.51 ± 0.01	102

Gravel Pits sl, 4—16% SM		25	0.39 ± 0.05**	102
Gravel Pits sl, 4—12% SM		30	0.54 ± 0.004	102
Gravel Pits sl, 4.8—10.7% SM		25	0.73 ± 0.07**	102
8.5—25°C	54.5			26
8.5—25°C	69.9			26
13.2—32.1°C	61.5			26
13.2—32.1°C	36.1			26

[a] SM = soil moisture content on a weight-per-weight basis.

[b] SMT = soil moisture tension content of 33 kPa, assuming field capacity is approximately equal to 33 kPa.

[c] exponential b in soil-moisture equation $k_s = A\ SM^b$.

[d] *,**, and ns = 95, 99%, and not significant confidence level, respectively.

[e] Questionable.

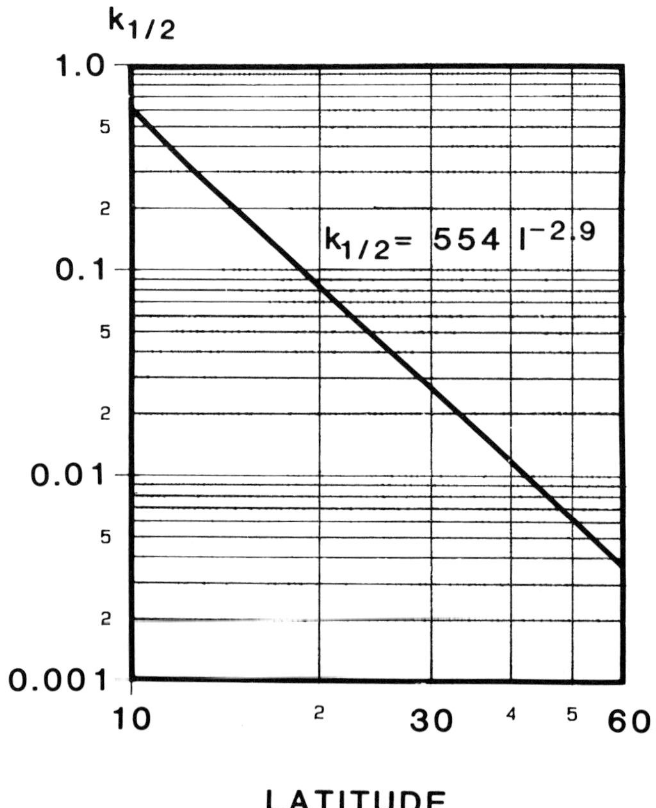

FIGURE 7.   Relationship between latitude and half-order rate constants $(k_{1/2})$ for the dissipation of picloram from soils at several locations in the United States and Canada. The initial $k_{1/2}$ values were calculated from:

$$k_{1/2} = 2(C_o^{0.5} - C_d^{0.5})(d)^{-1}$$

where C_o and C_d = initial and final picloram amounts. Hence C_d for any day and latitude can be estimated from $C_d = (C_o^{0.5} - d\,k_{1/2})(2)^{-1}$. (Data obtained from Hanaker, J. W., *in Organic Chemicals in the Soil Environment*, Goring, C. A. A. and Hamaker, J. W., Eds., Marcel Dekker, New York, 1972, 253.)

soil.[12,14] For many herbicides, volatilization becomes a more predominant dissipating pathway from soil surfaces compared with rhizosphere soil. Further, much greater environmental variations in temperature, wind run, soil moisture, solar radiation, and relative humidity occur on soil surface than in rhizosphere soil. The importance of including the effects of the environmental variable in any pesticide dissipation equation from soil surfaces is manifest, especially for prediction purposes. The approach of Walker[111] might be applicable for herbicide dissipation from soil surfaces, and Nash[36] has incorporated several soil and climatic variables in regression equations to determine their significance in describing disspation.

F. Soil Rhizosphere

Several herbicides are incorporated into the soil or can be incorporated into soil upon tillage after a soil surface or plant application. Dissipation through volatilization is greatly reduced when the herbicide is incorporated;[12] in fact, many herbicides are specifically incorporated to reduce their loss through volatilization (e.g., certain dinitroanilines and

Table 8
EFFECT OF SOIL pH UPON
DISAPPEARANCE TIMES (DAYS)
FOR SELECTED HERBICIDES

Herbicide	pH	$C_{1/2}$ (days)	k_s (per day)
Amitrole	6.0	1.4	0.495
Amitrole	7.0	1.6	0.433
Amitrole	8.0	1.3	0.533
Amitrole	5.3	(92)	0.0075
Amitrole	6.5	36	0.0193
Amitrole	7.5	56	0.124
Chloramben	4.3	36	0.0193
Chloramben	5.3	38	0.0182
Chloramben	6.5	41	0.0169
Chloramben	7.5	20	0.0347
2,4-D	4.3	20	0.0346
2,4-D	5.3	9.5	0.073
2,4-D	6.5	29	0.0239
2,4-D	7.5	<7	>0.099[a]
Dicamba	4.3	59	0.0117
Dicamba	5.3	19	0.036
Dicamba	6.5	17	0.041

[a] Allan E. Smith, personal communication for northern prairie soils.

Adapted from Hamaker, J. W., in *Organic Chemicals
in the Soil Environment,* Goring, C. A. A. and Hamaker,
J. W., Eds., Marcel Dekker, New York, 1972, 253.

thiocarboamates). Biodegradation usually is considered a more predominant pathway of dissipation for pesticides in the soil rhizosphere and usually is dependent upon the molecular structure of the herbicide. In general, microbial populations are highest near the soil surface and decrease exponentially with depth.

Much of the data available on herbicide dissipation from rhizosphere soil can be expressed mathematically by the pseudo first-order equation. This holds true especially for the more persistent herbicides.[3]

G. Subsoil

Little is known about herbicide dissipation from subsoil. It is generally assumed that dissipation would be less rapid in subsoil than in the rhizosphere or on the soil surface (although linuron degraded as rapidly at 30 to 60 cm in depth as at 0 to 10 cm and metribuzin nearly so[112]). The herbicide in subsoil is more likely to transform chemically rather than to degrade completely. Hydrolysis presumably occurs, and possibly some oxidation and reduction. Biological degradation presumably decreases rapidly below the rhizosphere, because of low microbial populations and anaerobic conditions.

Some transport through the subsoil is known to occur for some herbicides.[113] Presumably this occurs by both diffusion (short range) and leaching (long range), and presumably leaching becomes more dominant with depth because organic matter content decreases rapidly with depth below the rhizosphere. The rate of herbicide diffusion in subsoil under nonleaching conditions is not presently known, but is considered slow compared with leaching, except for high vapor-pressure herbicides.

H. Estimating Herbicide Dissipation

One could measure the concentration of a herbicide in or on soil with time and then develop an equation that would describe its dissipation rate. Measuring is the most common, and probably most accurate, way to develop an equation that can be used to estimate dissipation. As indicated in Section II. C., the dissipation of most herbicides from soils, especially if incorporated (Section II. F.), follows pseudo first-order kinetics. Therefore, it is possible to find a dissipation rate (k_s) from the literature (or Tables 3 to 8) to estimate the dissipation of a herbicide.

Nash[3] devised a scheme for collecting soil and environmental parameters from literature reporting pesticide dissipation. This information has been stored in a computer and contains 1100 items.[29] The file is available for use by anyone. The significance of this information allows one to rapidly obtain and compare the experimental information to his own situation.

However, the k_s values found from pseudo first-order kinetics are, as are all other equations developed to describe dissipation, valid only for the data set from which they were developed. This is not adequate. Efforts have therefore been made (1) to develop estimation equations from empirical relationships of the physiochemical characteristics of the herbicide and soil or environmental factors and (2) to estimate dissipation based strictly on the physiocochemical characteristics of the herbicide. The k_s value has been derived experimentally by several methods. Walker[102-104,111] demonstrated that nonvolatile herbicide dissipation from both surface and rhizosphere soils followed pseudo first-order kinetics and that this dissipation depended upon temperature and soil moisture. Walker and collaborators[43,85,94,102-104,111] determined herbicide dissipation at two or more temperatures under laboratory conditions and found $C_{1/2}$ values (k_s rather than $C_{1/2}$ values were used for this paper) from pseudo first-order equations. The activation energy (ΔE) from the Arrhenius equation was determined:

$$k_s = k_o\, e^{-\Delta E/RT} \tag{11}$$

where k_s = dissipation rate, k_o = frequency factor, R = 8.315 J mol^{-1}, and t = °K. Then, by knowing the ΔE and the K_{s1} value at T_1, a k_{s2} value can be estimated at T_2. For example, in Table 7 it was found that ΔE for asulam in Regina clay over a temperature range of 10 to 35°C was 46.8 kJ mol^{-1}. From Table 5, the k_{s1} at 10°C and 20% soil moisture content = 0.0141. Modifying Equation 11 to:

$$k_{s1} = k_{s2}\, e^{-\Delta E/R(1/T_1 - 1/T_2)} \tag{12}$$

where T_1 and T_2 = 10 and 35°C, respectively; k_{s2} = 0.0705 at 35°C, which is close to the k_s of 0.0774 reported in Table 5. Other k_s values can be calculated with equal confidence from these data.

Walker[112] found that the effect of soil moisture (SM) on herbicide persistence could be empirically derived:

$$C_{1/2} = a\, SM^{-b}$$

or

$$k_s = A\, SM^B \tag{13}$$

where a and b are constants. Like the Arrhenius equation, a k_s value at a different soil moisture can be calculated from a modification of Equation 13:

$$k_{s2} = k_{s1}\, \frac{SM_2^B}{SM_1^B} \tag{14}$$

(i.e., for asulam from Table 5: k_{s1} = 0.044 at 20% SM and 20°C, from Table 7: b = 1.45; the calculated k_{s2} = 0.1202 at 40% SM and 20°C, which is close to the k_s value of 0.1127 value given in Table 5.)

Further, by substituting Equation 14 into 12, a new k_s value can be estimated for any soil moisture content and temperature (i.e., asulam k_1 = 0.0261 at 20% SM and 15°C [Table 5]; b = 1.02, and ΔE = 40.1 kJ mol^{-1} in Table 7; k_2 = 0.121 at 40% SM and 30°C, which is close to the given k_s value of 0.1174 in Table 5).

Subsequently, by computer programming and using weather data to ascertain soil moisture and temperature, Walker and collaborators[43,85,102-104,111, 112] have simulated dissipation of several herbicides from both surface and rhizosphere soil quite accurately, but in other cases (possibly those that have greater daily soil moisture and temperature fluctuations or with higher organic matter content) the simulation usually underestimated herbicide dissipation.[85] Therefore, it would appear that, for several herbicides, data from carefully conducted laboratory experiments under several soil temperature and moisture regimes could be used with confidence to estimate or describe pesticide dissipation in the field. Table 7 gives the activation energy value (ΔE) and soil-moisture value (b), which can be inserted into Equations 12 and 14, respectively, to determine dissipation of several herbicides from soil under various temperature and soil-moisture conditions.

When dissipation is primarily by volatilization, several theoretical and empirical equations based upon the herbicide physicochemical properties, and usually some of the soil or environmental properties, may be used. These several methods have been reviewed and will not be repeated here.[114] Several of these were developed for herbicide volatilization from rhizosphere soil and hence require some knowledge about the diffusion of the herbicide in or from the soil.[24,25,115]

Hartley[116] provides an equation to determine flux in air of volatile herbicides and a simpler equation for less volatile herbicides. Both equations appear quite useful as a first approximation for both rhizosphere and surface soil, but in many cases neither the physicochemical nor environmental characteristics are available. Therefore, given the initial concentration and the estimated daily flux, a k_a value could be obtained from Equation 1, assuming pseudo first-order kinetics. Yet several data sets[5,10,12,14,16] have demonstrated that flux decline in air above a treated soil or crop apparently follows a series of first-order equation declines, which often can be described by Equation 4, or probably by a two-compartment model (Equation 5.)[8] The k_a in Equation 4 should approximate the k_s value of Equation 1. Hence, the herbicide dissipation from soil (Equation 1) can be estimated directly from the k_a of Equation 4, when dissipation is primarily by volatilization.

Hamaker[117] provides equations that can give an estimated daily flux, also. Like the Hartley equations,[116] the Hamaker equations show some potential for estimating (from flux as above using Equation 4) pesticide dissipation (k_s) from soil. Mayer et al.[118] have provided five models for determining pesticide flux. The Mayer et al. models are quite comprehensive, requiring considerable knowledge about the soil and environmental properties in addition to those of the herbicide. The Mayer et al. equations appear to provide greater sensitivity, but require more accurately described input parameters. Thus, if the parameters selected are inaccurate, the results could be off further than with less sensitive equations.

Jury et al.[24,25,115] provide models for estimating dissipation from volatilization plus degradation, volatilization, and leaching and their effective $C_{1/2}$ times at 1- and 10-cm soil depths. Utilization of these models requires only the soil K_{oc}, the pesticide K_h, and the net pesticide degradation rate k_b along with a set of standard properties to "classify and screen organic chemicals for their relative suceptibility to different loss pathways" (volatilization, leaching, degradation). Though the Jury et al. models appear quite applicable for screening purposes, they, like other models, lack (without further modification) the capability to estimate herbicide dissipation under various field conditions.

The major disadvantage of the above Hartley, Hamaker, Mayer, and Jury models for estimation purposes is that detailed pesticide physicochemical characteristics and environmental parameters are required and usually are not available.

Ideally, we would like to estimate pesticide dissipation (or volatilization) from soil surfaces with presently known pesticide physiochemical properties, environmental properties, and dissipation data sets. An equation for volatilization from nonadsorbing surfaces was given by Hartley:[116]

$$F_{b/a} = k_{sb} = \frac{VP_b(MW_b)^{0.5}}{VP_a(MW_a)^{0.5}} \times k_{sa} \tag{15}$$

where F = flux, VP = vapor pressure, and MW = molecular weight. The subscripts a and b denote the reference and subject compound, respectively. This equation might apply to moist surface soil in which adsorption would be minimal. Perhaps an improvement is an empirically derived equation (16-48 and 16-49)[114] incorporating the adsorption coefficient normalized on soil organic carbon (k_{oc}), and water solubility (K_w), in addition to VP:

$$k_s = 325 \frac{VP}{K_{oc}K_w} \text{ day}^{-1} \tag{16}$$

By using selected data from Nash,[14] Tables 9 and 10, and incorporating elements from both Equations 15 and 16, the following relationship was found:

$$k_s = 3.39[H(K_{oc})^{-1}]^{0.413+0.047} \qquad (r^2 = 0.80),\ n = 21 \tag{17}$$

where H = Henry's law constant (Pa m^3 mol^{-1}). Equation 17 now incorporates important physiochemical properties, which are available for many pesticides and are the same physicochemical properties selected by Jury et al.[24] in developing their model. Obviously K_{oc} exerts considerable influence on both Equation 16 and 17, as well as the Jury et al. equation, and should be determined or selected with care. K_{oc} can be obtained, respectively, from K_{ow} (octanol/water partitioning coefficient) and K_w by:[119-121]

$$K_{oc} = 0.63\ K_{ow} = 0.48\ K_{ow}$$

$$= 18{,}750\ K_w^{-0.686}$$

$$= 4365\ K_w^{-0.55} \tag{18}$$

Equations 16 and 17 were tested on both laboratory and field data sets and gave reasonably good results (Tables 9, 10).[122-136]

Equation 17 does not take into consideration any of the environmental variables. Now, however, with an estimated k_s value at 25°C, Equation 1 can be modified to include temperature.[134,135] The k_s (Equation 1) was divided by 25 to obtain a time/temperature dissipation rate ($-k_{st}$) at 25°C.

$$C_d = C_o\ e^{-k_{st}d_d} \tag{19}$$

where d_d = degree − days or cumulative daily mean temperature above 0°C for the time in question. The goodness-of-fit for values calculated from Equation 19 vs. the observed values found in Table 5, gave y = 0.93× with $r^2 = 0.91$ and n = 54. (Air temperatures often range as much as 20°C lower than bare soil surface temperatures on bright sunny days.)

Table 9

PESTICIDE DISSIPATION FROM LABORATORY SOIL SURFACES AS RELATED TO THE PESTICIDE PHYSICOCHEMICAL PROPERTIES AT 25°C

Pesticide	Molecular weight (MW) (g)	Vapor pressure (VP) (mPa)	Water solubility (S) ($\mu g\ g^{-1}$)	Soil adsorption coefficient (K_{oc}) $\left[\dfrac{\mu g\ P(gA)^{-1}}{\mu g\ P(gW)^{-1}}\right]$ [a]	Dissipation rate (k_s—per day) Measured	Estimated from Equation 16	15[b]	17
Atrazine	215.7	0.113	33	150	0.015	0.0074	0.0036	0.022
Carbofuran	221.3	0.27	415	10	0.029	0.021	0.0087	0.034
Chlorpyrifos	350.5	2.53	2	13,000	0.23	0.032	0.10	0.049
DDT	354.5	0.036	0.0012	240,000	0.017	0.041	0.0015	0.054
Dinoseb	240.2	6.67	50	700	0.027	0.062	0.22	0.055
Diuron	233.1	0.25	42	150	0.058	0.013	0.0083	0.028
Lindane	290.8	10.67	7	2,500	1.00	0.20	0.40	0.10
Nitrapyrin	230.9	400	40	600	13.86	5.4	13.22	0.34
Trifluralin	335.0	17.33	0.6	6,000	0.69	1.6	0.69	0.24

[a]   p = pesticide, A = adsorbent, W = water.
[b]   Trifluralin was the reference pesticide.

Adapted from Thomas, R. G., in *Handbook of Chemical Property Estimation Methods*, Lyman, W.J., Rechl, W. F., and Rosenblatt, D.H., Eds., McGraw-Hill, New York,1982, chap. 2.

Table 10
PESTICIDE DISSIPATION FROM FIELD AND CHAMBER SOIL SURFACES AS RELATED TO THE PESTICIDE PHYSICOCHEMICAL PROPERTIES AT 25°C

| | | | | Soil adsorption coefficient (K_{oc}) $\left[\dfrac{\mu g\ P(gA)^{-1}}{\mu g\ P(gW)^{-1}}\right]$ [a] | Dissipation rate (k_s — per day) | | | Ref. | | | |
| | Molecular weight (MW) (g) | Vapor pressure (VP) (mPa) | Water solubility (S) ($\mu g\ g^{-1}$) | | | Estimated from Equation | | | | | |
Pesticide					Measured	16	17	VP	S	K_{oc}	Measured
Alachlor	269.8	2.94	242	200	0.077	0.020	0.036	122	122	121	40
Butylate	217.4	1733	45	540	3.6	23	0.610	122	121	121	12
Dieldrin	380.9	0.8	0.25	35,600	0.055	0.034	0.049	124	124	121	14
Endrin	380.9	0.7[b]	0.23	34,000	0.036	0.034	0.048		121	121	14
Fluchloralin	355.7	3.53	1.0	4,000	0.099—0.13	0.29	0.120	122	130	121	35
Fluometuron	232.2	0.067	90	200	0.023—0.043	0.0012	0.011	122	122	121	109
Fluridone	329.3	0.01	12	40	0.0041	0.0067	0.025	122	122	121	95
Heptachlor	373.4	25	0.3	30,000	0.28	1.0	0.20	125	122	121	14
Lindane	290.8	10.7	7	2,500	0.16	0.20	0.10	126	121	121	14
Malathion[c]	330.3	4	145	280	0.018—0.064	0.032	0.048	127	127	121	134
Methyl parathion[d]	263.21	2.29	60	424	0.010—0.034	0.029	0.042	128	127	131	135
2,3,7,8-TCDD	322	2×10^{-4}	3.2×10^{-4}	800,000	0.0077	0.00026	0.0064	132	132	132	5
Toxaphene	414	0.13	0.4	98,600	0.010	0.0011	0.013	127	127	133	136
Trifluralin	335.0	17.33	0.6	6,000	0.69	1.6	0.24	122	121	121	14

[a]   P = pesticide, A = adsorbant, W = water.

[b]   Estimated.

[c]   8 to 12°C.

[d]   Laboratory.

Table 11
AVAILABILITY (A) CLASSES
FOR PESTICIDES IN SOIL
CONTAINING 15% MOISTURE

A	K_d
1	<0.05
2	0.15
3	0.45
4	1.35
5	2.85
6	5.85
7	14.9
8	29.9
9	59.9
10	150
11	300
12	600
13	1500
14	>1500

Adapted from Briggs, G. G., in *Proc. British Crop Protection Council Symp.: The Persistence of Insecticides and Herbicides,* Monogr. 17, British Crop Protection Council, London, 1976, 41.

Perhaps similar adjustments could be made in the Jury et al. models[24,25,115] to make them more useful for field estimation purposes.

When the soil dries, pesticide volatilization decreases sharply.[9,129] The soil moisture content at which volatilization drops dramatically has been found to be equivalent to a monomolecular layer of water covering the soil surface. This soil moisture content (%) can be calculated from the soil specific-surface equation ($SSSE = m^2\ g^{-1}$),[22] where the monomolecular water thickness is 3×10^{-8} cm:[129]

$$SSSE = 10(\%\ OC) + \{[A(100 - M) + (7.5 * M)]/100\}(\%c) \qquad (20)$$
$$+ \ 0.04(\%si) + 0.0005(\%s)$$

where A = percent of clay as kaolinite, 0.07 to 0.3 (default 0.25) or illite, 0.65 to 1 (default 0.9); M = percent of clay as montmorillonite and vermiculite, 6 to 8 (default 7.5); and OC = organic carbon (OC = 0.58 [OM]).

Briggs[137] proposed a method of predicting pesticide half-concentration time ($C_{1/2}$) in soil from the physicochemical properties of the pesticide and soil, primarily the former. He suggested that the two criteria needed for predicting pesticide persistence (or, here, dissipation [k_s]) are the *availability* (in soil solution) of the pesticide and its *degradability* (functional group lability). The product of these two criteria, availability (A) and degradability (D), estimates the $C_{1/2}$ of the pesticide in weeks. The $C_{1/2}$ (weeks) can be converted to a k_s (per day) by

$$k_s = \frac{0.693}{C_{1/2}(7)} \qquad (21)$$

Briggs[137] established 14 availability (A) classes (Table 11) from K_d values by

Molecule	Stability in Soil
RCOOR′	1—3
RCN	1—5
RCONH$_2$	1—5
RCONHR′	1—5
RCH=CHR	2—4
RCH$_2$OH	1—2
ArOH	1—8
ArNH$_2$	1—8
ArNO$_2$	3—4
ArCOOH	1—10
RCH=CH$_2$Br	1—1
RCH$_2$Br	3—3
RCH=CHCH$_2$Cl	3—3
RCH$_2$Cl	5—5
RCH=CHCl	6—6
ArCH$_3$	4—4
ArCl	5—5
ArCl$_2$	7—7
ArCl$_4$	10—10
RCCl$_3$	4—5
ArNHCOOR	3—4
RNHCOOAr	5—6
ArNHCONR$_2$	7—8
ArNHR	2—9
ArOCH$_3$	3—6
ROCH$_3$	2—3
RSCH$_3$	1—2
OPO(OCH$_3$)$_2$	3—3
RPS(OCH$_2$CH$_3$)$_2$	4—4
OPS(OCH$_3$)$_2$	4—4
OPO(OCH$_2$CH$_3$)$_2$	5—5
OPS(OCH$_2$CH$_3$)$_2$	6—6
SPS(OCH$_2$CH$_3$)$_2$	7—7

FIGURE 8. Degradability of functional groups on a pesticide molecule, based on a 1 to 10 scale. (Adapted from Briggs, G. G., in *Proc. British Crop Protection Council Symp.: The Persistence of Insecticides and Herbicides*, Monogr. 17, British Crop Protection Council, London, 1976, 41.)

$$K_d = (\% \ OM \times K_{om})(100)^{-1} \tag{22}$$

where OM and K_{om} = soil organic matter and soil OM/water distribution coefficient. K_{om} was determined from the octanol/water distribution coefficient (k_{ow}) by

$$K_{om} = 4.17 \ K_{ow}^{0.52} \tag{23}$$

Briggs[137] established ten degradability (D) classes based on the pesticide functional groups (Figure 8) (further information on this approach is contained in Brigg's excellent discussion). A readily degraded functional group on a pesticide is assigned a low number on a scale of 1 to 10, whereas a stable functional group is assigned a high number.

Many functional groups may have variable degradability because of the steric and electronic effects in the rest of the molecule. Steric nonhindered esters (RCOOR′) are very short lived in soil, for example, and are assigned a degradability of 1, whereas steric hindered esters are more stable and are assigned a degradability of 3.

Persistence of aromatic carboxylic acids (ArCOOH) depends highly on the other substituents in the aromatic ring. Benzoic acid is degraded in soil in a few hours, but 2,3,6-trichlorobenzoic acid (TBA) is very stable in soil because of both steric and electronic effects of the chlorine substituents. Hence, benzoic would be assigned a degradability value of 1, and TBA, 10

When Briggs[137] compared his AD values to those obtained from the literature, the values tended to underestimate the persistence of long-lived pesticides and overestimate the persistence of short-lived pesticides, being quite poor for the latter. It would appear that further refinement of the method is required before it can be considered reliable.

In summary, herbicide dissipation from soil surfaces appears to follow at least two distinct "pseudo" first-order rates, i.e., an initial fast rate followed by a slower rate. Estimating herbicide dissipation during the initial period can be done by using a combination of the herbicide physicochemical characteristics and certain environmental parameters (soil temperature and moisture), when dissipation is primarily by volatilization. Estimating herbicide dissipation during the second slower phase is more formidable without measuring, but the approach of Briggs might be useful.

One should keep in mind that none of these methods has received extensive testing. Thus, they may not be valid for conditions other than those for which the methods were developed or tested. This holds true especially for the empirically derived equations. The Hartley, Hamaker, Mayer et al. and Jury et al. equations all suffer the same shortcomings of the empirically derived equations in that soil (i.e., temperature and moisture), and environmental parameters are not adequately tested over a wide range of conditions.[23,24,115-118]

I. Carry-Over Aspects and Problems

Numerous examples can be cited where herbicides have been phytotoxic to subsequent crops either during the same season or in following seasons. This was especially true prior to 1970 when testing potential herbicides was less extensive than it is now. Carry-over problems (at least from season to season) are more likely to occur in colder climates than in warm climates (Figures 6, 7, Tables 5, 7).

Herbicide carry-over can be calculated by using Equation 1 or subsequent modified versions if a k_s value is known. Further, under repeated applications in which the herbicide is never completely dissipated, a residual of the herbicide will build up and can be calculated from:

$$CO_n = UA(R^n) + CO_{n-1}(+ UA)_{for\ max} \tag{24}$$

where CO_n = carry-over at the nth application, UA = units applied, and R = the ratio between the minimum residual and maximum amount. This has been depicted by Kearney et al.[30] in Figure 9. Notice a maximum for the residuals will eventually be reached and is dependent upon the R value in Equation 24. When four units of applied herbicide decreased to 0.8 units (80% dissipated) prior to the next application, the number of applications to reach maximum residual level would be five, but for all practical purposes, the maximum level would be reached after the third application.

III. SUMMARY AND CONCLUSION

Dissipation of pesticides on surface soil, in rhizosphere soil, and, presumbly, in subsoil usually follows a simple exponential equation, which is a lumped sum of all the loss pathways and the chemical and environmental factors affecting dissipation. An equation developed from a given data set theoretically is valid only for that data set and should not be used to estimate pesticide dissipation under different conditions without additional testing and cal-

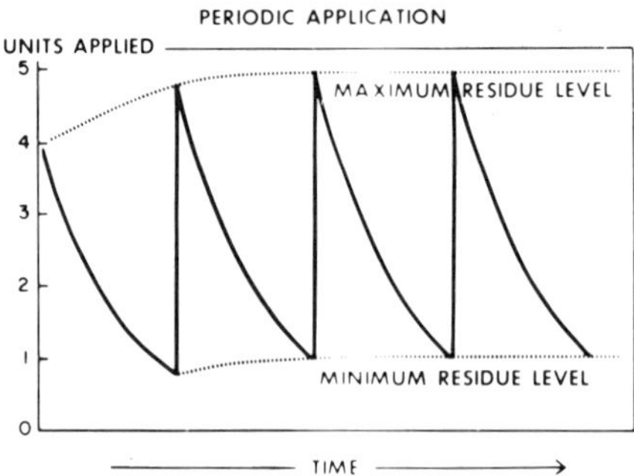

FIGURE 9.   Maximum and residual herbicide levels from repeated applications and in which herbicide amounts did not reach zero prior to the next application. (From Kearney, P. C., Nash, R. G., and Isensee, A. R., in *Chemical Fallout: Current Research on Persistent Pesticides,* Miller, M. W. and Berg, G. G., Eds., Charles C. Thomas, Springfield, Illinois, 1969. With permission.)

ibration. If there are no alternatives, or additional means to test or validate the equation, its application must be considered with a great deal of caution.

Dissipation activation energies (ΔE) (Arrhenius equation) have been determined for several pesticides. The activation energies can be used to estimate pesticide dissipation under conditions like those from which the dissipation activation energies were determined, but should be used only with caution under different conditions. Similar to the dissipation activation energy, equations describing soil moisture influence on dissipation have been developed for given situations. These can be used under like conditions, but should be used with caution to estimate dissipation under different conditions.

Exponential integral-variable equations are valid only for the data sets from which they were developed, but they may be useful to estimate pesticide dissipation under different conditions, provided caution is taken, as described above. Soil moisture and soil or air temperatures were the environmental variables most often associated with the integral-variable equations, and soil moisture and temperature are the two variables addressed in the Arrhenius and soil-moisture influence equations. The advantage of either Arrhenius plus soil-moisture influence or integral-variable equations over exponential equations is that they use environmental parameters that directly affect dissipation.

A conceptual approach (Equation 10), in which the dissipation rate from each loss pathway is defined and in which environmental influence are reflected, has the greatest potential because true kinetics can be assigned. Unfortunately, separating the several pathways to measure a dissipation pathway accurately under precisely measured environmental (experimental) conditions is rarely obtainable.

When pesticide dissipation is primarily by volatilization, several theoretical and empirical equations have been derived to estimate pesticide flux. Subsequently, the dissipation rate (k_s) can be estimated from the flux and original amount. Some empirically derived equations (which include some of the pesticide physiocochemical characteristics and environmental parameters) appear to be capable of estimating pesticide dissipation from soil surfaces. For example, the Jury et al. models[24,26,115] should have considerable potential for estimating dissipation when volatilization is a major pathway.

All of the empirically derived equations are only valid for the data sets from which they were developed. Likewise, many of the theoretical equations may be valid only for the data sets from which they were tested, because assumptions are always made. In principle, however, the theoretical equations based upon sound thermodynamics and kinetics should prove the most useful in the long run, despite their present shortcomings.

IV. SUGGESTIONS FOR FUTURE RESEARCH

Research on herbicide dissipation should be continued, with emphasis upon

1. Identifying, measuring, and cataloging all soil and climatic factors that affect dissipation.
2. Identifying and measuring major and minor loss pathways.
3. Obtaining better physicochemical properties of the herbicide.

Researchers should be cognizant of the needs of modelers and hence, should plan their research to provide the environmental properties necessary for a true kinetic approach to pesticide dissipation from soil.

REFERENCES

1. **Weber, J. B., Monaco, T. J., and Worsham, A. D.,** What happens to herbicides in the environment?, *Weeds Today,* 4, 16, 1973.
2. **Edwards, C. A.,** Insecticide residues in soils, *Residue Rev.,* 13, 83, 1966.
3. **Nash, R. G.,** Dissipation rate of pesticides from soils, in CREAMS: Field-Scale Model for Chemicals, Runoff, and Erosion from Agricultural Management Systems, Vol. 3, Knisel, W. G., Ed., U.S. Government Printing Office, Washington, D. C., 1980, 560.
4. **Seiber, J. N. and Woodrow, J. E.,** Methods for studying pesticide atmospheric dispersal and fate at treated areas, *Residue Rev.,* 85, 217, 1983.
5. **Nash, R. G. and Beall, M. L., Jr.,** Distribution of silvex, 2,4-D, and TCDD applied to turf in chambers and field plots, *J. Agric. Food Chem.,* 28, 614, 1980.
6. **Brewer, F., Talbert, R. E., and Lavy, T. L.,** Effect of DBCP on fluchloralin persistence, *Weed Sci.,* 29, 605, 1981.
7. **Hill, B. D., Charnetski, W. A., Schaalje, G. B., and Schaber, B. D.,** Persistence of fenvalerate in alfalfa: effect of growth dilution and heat units on residue half-life, *J. Agric. Food Chem.,* 30, 653, 1982.
8. **Hill, B. D. and Schaalje, G. B.,** A two-compartment model for the dissipation of deltamethrin on soil, *J. Agric. Food Chem.,* 33, 1001, 1985.
9. **Spencer, W. F. and Cliath, M. M.,** Volatility of DDT and related compounds, *J. Agric. Food Chem.,* 20, 645, 1972.
10. **Taylor, A. W.,** Post-application volatilization of pesticides under field conditions, *J. Air Pollut. Control Assoc.,* 28, 922, 1978.
11. **Williams, J. H. and Eagle, D. J.,** Persistence of dichlorben in a sandy soil and effects of resides on plant growth, *Weed Res.,* 19, 315, 1979.
12. **Nash, R. G.,** Distribution of butylate, heptachlor, lindane, and dieldrin emulsifiable concentrated and butylated microencapsulated formulations in microagroecosystem chambers, *J. Agric. Food Chem.,* 31, 1195, 1983.
13. **Willis, G. H., McDowell, L. L., Harper, L. A., Southwick, L. M., and Smith, S.,** Seasonal disappearance and voltilization of toxaphene and DDT from a cotton field, *J. Environ. Qual.,* 12, 80, 1983.
14. **Nash, R. G.,** Comparative volatilization and dissipation rates of several pesticides from soil, *J. Agric. Food Chem.,* 31, 210, 1983.
15. **Nash, R. G. and Beall, M. L., Jr.,** Fate of maneb and zineb fungicides in microagroecosystem chambers, *J. Agric. Food Chem.,* 28, 322, 1980.
16. **Nash, R. G., Beall, M. L., Jr., and Harris, W. G.,** Toxaphene and 1,1,1-trichloro-2,2-bis(*p*-chlorophenyl)ethane (DDT) losses from cotton in an agroecosystem chamber, *J. Agric. Food Chem.,* 25, 336, 1977.
17. **Gile, J. D.,** 2,4-D — Its distribution and effects in a ryegrass ecosystem, *J. Environ. Qual.,* 12, 406, 1983.
18. **Gile, J. D. and Gillet, J. W.,** Fate of selected fungicides in a terrestrial laboratory ecosystem, *J. Agric. Food Chem.,* 27, 1159, 1979.
19. **Gile, J. D., Collins, J. C., and Gillet, J. W.,** Fate of selected herbicides in a terrestrial laboratory microcosm, *Environ. Sci. Technol.,* 14, 1124, 1980.
20. **Glotfelty, D. E.,** The atmosphere as a sink for applied pesticides, *J. Air Pollut. Control Assoc.,* 29, 917, 1978.

21. **Green, R. E. and Karickhoff, S. W.,** Estimation of pesticide sorption coefficients for soils and sediments, in Supporting Papers for "SWAM", Vol. 3, Agricultural Research Service, U.S. Department of Agriculture, Washington, D.C., in press.

22. **Pionke, H. B. and DeAngelis, R. J.,** Method for distributing pesticide loss in field runoff between the solution and adsorbed phase, in CREAMS: Field-Scale Model for Chemicals, Runoff, and Erosion from Agricultural Management Systems, Vol. 3, Knisel, W. G., Ed., U. S. Government Printing Office, Washington, D. C., 1980, 607.

23. **Hartley, G. S. and Graham-Bryce, I. J.,** *Physical Principles of Pesticide Behavior,* Vol. I and II, Academic Press, New York, 1980.

24. **Jury, W. A., Spencer, W. F., and Farmer, W. J.,** Behavior assessment model for trace organics in soil. I. Model description, *J. Environ. Qual.,* 12, 558, 1983.

25. **Jury, W. A., Spencer, W. F., and Farmer, W. J.,** Behavior assessment model for trace organics in soil. III. Application of screening model, *J. Environ. Qual.,* 13, 573, 1984.

26. **Hamaker, J. W.,** Decomposition: quantitative aspects, in *Organic Chemicals in the Soil Environment,* Goring, C. A. A. and Hamaker, J. W., Eds., Marcel Dekker, New York, 1972, 253.

27. **Leistra, M.,** Evaluation of rate coefficients for consecutive reactions of pesticides in soil, *J. Environ. Sci. Health,* B13, 343, 1978.

28. **Soulas, G.,** Mathematical model for microbial degradation of pesticides in the soil, *Soil Biol. Biochem.,* 14, 107, 1982.

29. **Nash, R. G. and Osborne, E. M.,** Environmental Chemical Dissipation (ECD) File, First Annu. Meet. Soc. Environmental Toxicology and Chemistry, November 23 to 25, 1980.

30. **Kearney, P. C., Nash, R. G., and Isensee, A. R.,** Persistence of pesticide residues in soils, in *Chemical Fallout: Current Research on Persistent Pesticides,* Miller, M. W. and Berg, G. G., Eds., Charles C Thomas, Springfield, Illinois, 1969.

31. **Obrigawitch, T., Martin, A. R., and Roeth, F. W.,** The influence of temperature, moisture, and prior EPTC application on the degradation of EPTC in soils, *Weed Sci.,* 30, 175, 1982.

32. **Walker, A.,** A simulation model for prediction of herbicide persistence, *J. Environ. Qual.,* 3, 396, 1974.

33. **Nigg, H. N., Allen, J. C., Brooks, R. F., Edwards. G. J., Thompson, N. P., King, R. W., and Blagg, A. H.,** Dislodgeable residues of ethion in Florida citrus and relationships to weather variables, *Arch. Environ. Contam. Toxicol.,* 6, 257, 1977.

34. **Nash, R. G.,** Integrated and nonintegrated equations for describing pesticide dissipation from soil and air, *Am. Soc. Agron. Abstr.,* p. 36, 1983.

35. **Savage, K. E. and Jordan, T. N.,** Persistence of three dinitroaniline herbicides on the soil surface, *Weed Sci.,* 28, 105, 1980.

36. **Nash, R. G.,** Methods for estimating pesticide dissipation from soils, in SWAM: Small Watershed Model, Vol. 3, DeCoursey, D., Ed., Agricultural Research Service, U.S. Department of Agriculture, Washington, D.C., unpublished.

37. **Dawson, J. H.,** Dodder control in alfalfa with dinoseb and D(−)3-chlorophenylcarbamoyloxy) 2N-isopropylpropionamide, *Weeds,* 5, 551, 1971.

38. **Kennedy, J. M. and Talbert, R. E.,** Comparative persistence of dinitroaniline type herbicides on the soil surface, *Weed Sci.,* 25, 373, 1977.

39. **Romanowski, R. R. and Libik, A. W.,** Soil persistence of isopropalin, nitralin, and trifluralin, *Weed Sci.,* 26, 258, 1978.

40. **Zimdahl, R. L. and Clark, S. K.,** Degradation of three acetanilide herbicides in soil, *Weed Sci.,* 30, 545, 1982.

41. **Wu, T. L.,** Dissipation of the herbicides atrazine and alachlor in a Maryland corn field, *J. Environ. Qual.,* 9, 459, 1980.

42. **Menges, R. M. and Tamez, S.,** Movement and persistence of bensulide and trifluralin in irrigated soil, *Weed Sci.,* 22, 67, 1974.

43. **Walker, A. and Zimdahl, R. L.,** Simulation of the persistence of atrazine, linuron and metolachlor in soil at different sites in the U.S.A., *Weed Res.,* 21, 255, 1981.

44. **Walker, A.,** Simulation of the persistence of eight soil-applied herbicides, *Weed Res.,* 18, 305, 1978.

45. **Smith, A. E. and Hayden, B. J.,** Comparison of the persistence of EPTC, metribuzin, and propanil in Saskatchewan field soils, *Bull. Environ. Contam. Toxicol.,* 29, 243, 1982.

46. **Smith, A. E.,** Degradation of dicamba in prairie soils, *Weed Res.,* 13, 373, 1973.

47. **Alton, J. D. and Stritzke, J. F.,** Degradation of dicamba, picloram, and four phenoxy herbicides in soils, *Weed Sci.,* 21, 556, 1973.

48. **Scifres, C. J. and Allen, T. J.,** Dissipation of dicamba from grassland soils of Texas, *Weed Sci.,* 21, 393, 1973.

49. **Ingram, G. H. and Pullin, E. M.,** Persistence of bromoxynil in three soil types, *Pestic. Sci.,* 5, 287, 1974.

50. **Beynon, K.I. and Wright, A. N.**, Persistence, penetration, and breakdown of chlorthiamid and dichlobenil herbicides in field soil of different types, *J. Sci. Food Agric.*, 19, 718, 1968.
51. **Fryer, J. D., Hance, R. J., and Ludwig, J. W.**, Long-term persistence of paraquat in a sandy loam soil, *Weed Res.*, 15, 189, 1975.
52. **Smith, A. E.**, Soil persistence studies with [¹⁴C]MCPA in combination with other herbicides and pesticides, *Weed Res.*, 22, 137, 1982.
53. **Walter, J. P., Eastin, E. F., and Merkle, M. G.**, The persistence and movement of fluorodifen in soils and plants, *Weed Res.*, 10, 165, 1970.
54. **Murty, A. S., Miles, J. R. W., and Tu, C. M.**, Persistence and mobility of nitrofen (niclofen, TOK (R)) in mineral and organic soils, *J. Environ. Sci. Health*, B17, 143, 1982.
55. **Jacques, G. L. and Harvey, R. G.**, Persistence of dinitroaniline herbicides in soil, *Weed Sci.*, 27, 660, 1979.
56. **Savage, K. E.**, Persistence of several dinitroaniline herbicides as affected by soil moisture, *Weed Sci.*, 26, 465, 1978.
57. **Poku, J. A. and Zimdahl, R. L.**, Soil persistence of dinitramine, *Weed Sci.*, 28, 650, 1980.
58. **Hayden, B. J. and Smith, A. E.**, Comparison of the persistence of ethalfluralin and trifluralin in Saskatchewan field soils, *Bull. Environ. Contam. Toxicol.*, 25, 508, 1980.
59. **Savage, K. E.**, Nitralin and trifluralin persistence in soil, *Weed Sci.*, 21, 285, 1973.
60. **Barrett, M. R. and Lavy, T. L.**, Effects of soil water content on oxadiazon dissipation, *Weed Sci.*, 32, 697, 1984.
61. **Zimdahl, R. L., Catizone, P., and Butcher, A. C.**, Degradation of pendimethalin in soil, *Weed Sci.*, 32, 408, 1984.
62. **Coffman, C. B. and Gentner, W. A.**, Persistence of several controlled release formulations of trifluralin in greenhouse and field, *Weed Sci.*, 28, 21, 1980.
63. **Golab, T., Althaus, W. A., and Wooten, H. L.**, Fate of [¹⁴C] trifluralin in soil, *J. Agric. Food Chem.*, 27, 163, 1979.
64. **Smith, A. E.**, Persistence of trifluralin in small field plots as analyzed by a rapid gas chromatographic method, *J. Agric. Food Chem.*, 20, 829, 1972.
65. **Smith, A. E. and Hayden, B. J.**, Field persistence studies with triallate and trifluralin both singly and in combination with chloramben, *Bull. Environ. Contam. Toxicol.*, 29, 240, 1982.
66. **Savage, K. E. and Barrentine, W. L.**, Trifluralin persistence as affected by depth of soil incorporation, *Weed Sci.*, 17, 349, 1969.
67. **White, A. W., Jr., Harper, L. A., Leonard, R. A., and Turnbull, J. W.**, Trifluralin volatilization losses from a soybean field, *J. Environ. Qual.*, 6, 105, 1977.
68. **Smith, A. E.**, Soil persistence experiments with [¹⁴C]2,4-D in herbicide mixtures and field persistence studies with triallate and trifluralin both singly and combined, *Weed Res.*, 19, 165, 1979.
69. **Smith, A. E. and Hayden, B. J.**, Relative persistence of MCPA, MCPB and mecoprop in Saskatchewan soils, and the identification of MCPA in MCPB treated soils, *Weed Res.*, 21, 179, 1981.
70. **Lutz, J. F., Byers, G. E., and Sheets, T. J.**, The persistence and movement of picloram and 2,4,5-T in soils, *J. Environ. Qual.*, 2, 485, 1973.
71. **Wilson, R. G.**, Accelerated degradation of thiocarbamate herbicides in soil with prior thiocarbamate herbicide exposure, *Weed Sci.*, 32, 264, 1984.
72. **Smith, A. E.**, Degradation, adsorption and volatility of diallate and triallate in prairie soils, *Weed Res.*, 10, 331, 1970.
73. **Obrigwitch, T., Roeth, F. W., Martin, A. R., and Wilson, R. G., Jr.**, Addition of R-33865 to EPTC for extended herbicide activity, *Weed Sci.*, 30, 417, 1982.
74. **Kells, J. J., Rieck, C. E., Blevins, R. L., and Muir, W. M.**, Atrazine dissipation as affected by surface pH and tillage, *Weed Sci.*, 28, 101, 1980.
75. **Kahn, S. U., Marriage, P. B., and Hammill, A. S.**, Effects of atrazine treatment of a corn-field using different application methods, times, additives on the persistence of residues in soil and their uptake by oat plants, *J. Agric. Food Chem.*, 29, 216, 1981.
76. **Ghadiri, H., Shea, P. J., Wicks, G. A., and Haderlie, L. C.**, Atrazine dissipation in conventional-till and no-till sorghum, *J. Environ. Qual.*, 13, 549, 1984.
77. **Hiltbold, A. E. and Buchanan, G. A.**, Influence of soil pH on persistence of atrazine in the field, *Weed Sci.*, 25, 515, 1977.
78. **Libik, A. W. and Romanowski, R. R.**, Soil persistence of atrazine and cyanazine, *Weed Sci.*, 24, 627, 1976.
79. **Bauman, T. T. and Ross, M. A.**, Effect of three tillage systems on the persistence of atrazine, *Weed Sci.*, 31, 423, 1983.
80. **Yoo, J. Y., Muir, D. C. G., and Baker, B. E.**, Persistence and movement of cyanazine and procyazine in soil under field conditions, *Can. J. Soil Sci.*, 61, 237, 1981.

81. **Savage, K. E.,** Metribuzin persistence in soil, *Weed Sci.,* 25, 55, 1977.
82. **Ladlie, J. S., Meggitt, W. F., and Penner, D.,** Role of pH on metribuzin dissipation in field soils, *Weed Sci.,* 24, 508, 1976.
83. **Banks, P. A. and Robinson, E. L.,** The influence of straw mulch on the soil reception and persistence of metribuzin, *Weed Sci.,* 30, 164, 1982.
84. **Walker, A.,** Simulation of herbicide persistence in soil. I. Simazine and prometryn, *Pestic. Sci.,* 7, 41, 1976.
85. **Walker, A., Hance, R. J., Allen, J. G., Briggs, G. G., Chen, Yuh-lin, Gaynor, J. D., Hogue, E. J., Malquori, K., Moody, K., Moyer, J. R., Pestemer, W., Rahman, A., Smith, A. E., Streibig, J. C., Torstensson, N. T. L., Widyanto, L. S., and Zandvoort, R.,** EWRS herbicide-soil working group: collaborative experiment on simazine persistence in soil, *Weed Res.,* 23, 373, 1983.
86. **Marriage, P. B., Khan, S. U., and Saidak, W. J.,** Persistence and movement of terbacil in peach orchard soil after repeated annual applications, *Weed Res.,* 17, 219, 1977.
87. **Koning, M., Sieberg, K., and Ladewig, C.,** Untersuchungen zur Persistenz von Lenacil im Boden, *Nachrichtenbl. Pflanzenschutz DDR,* 34, 1980, 138.
88. **Ragab, M. T. H., Everett, C. F., and Chaisson, C. A.,** Persistence and movement of chlorbromuron in potato soil, *J. Environ. Sci. Health,* B14, 181, 1979.
89. **McCormick, L. L. and Hiltbold, A. E.,** Microbial decomposition of atrazine and diuron in soil, *Weeds,* 14, 77, 1966.
90. **Khan, S. U., Marriage, P. B. and Saidak, W. J.,** Persistence and movement of diuron and 3,4-dichloroaniline in an orchard soil, *Weed Sci.,* 24, 583, 1976.
91. **Hance, R. J. and Haynes, R. A.,** The kinetics of linuron and metribuzin decomposition in soil using different laboratory systems, *Weed Res.,* 21, 87, 1981.
92. **Shadbolt, C. A., Whiting, F. L., and Lyons, J. M.,** Persistence of monuron and neburon under semi-arid field conditions, *Weeds,* 12, 304, 1964.
93. **Emmerich, W. E., Helmer, J. D., Renard, K. G., and Land, L. J.,** Fate and effectiveness of tebuthiuron applied to a rangeland watershed, *J. Environ. Qual.,* 13, 382, 1984.
94. **Walker, A. and Brown, P. A.,** Measurement and prediction of chlorsulfuron persistence in soil, *Bull. Environ. Contam. Toxicol.,* 30, 365, 1983.
95. **Banks, P. A., Ketchersid, M. L., and Merkle, M. G.,** The persistence of fluridone in various soils under field and controlled conditions, *Weed Sci.,* 27, 631, 1979.
96. **Rueppel, M. L., Brightwell, B. B., Schaefer, J., and Marvel, J. T.,** Metabolism and degradation of glyphosate in soil and water, *J. Agric. Food Chem.,* 25, 517, 1977.
97. **Barrett, M. R. and Lavy, T. L.,** Effects of soil water content on pendimethalin dissipation, *J. Environ. Qual.,* 12, 504, 1983.
98. **Fryer, J. D., Smith, P. D., and Ludwig, J. W.,** Long-term persistence of picloram in a sandy loam soil, *J. Environ. Qual.,* 8, 83, 1979.
99. **Keys, C. H. and Friesen, H. A.,** Persistence of picloram in soil, *Weed Sci.,* 16, 341, 1968.
100. **Smith, A. E. and Hsiao, A. I.,** Persistence studies with the herbicide sethoxydim in prairie soils, *Weed Res.,* 23, 253, 1983.
101. **Smith, A. E. and Walker, A.,** A quantative study of asulam in soil, *Pestic. Sci.,* 8, 449, 1977.
102. **Walker, A.,** Simulation of herbicide persistence in soil. II. Simazine and linuron in long-term experiments, *Pestic. Sci.,* 7, 50, 1976.
103. **Walker, A.,** Simulation of herbicide persistence in soil. I. Simazine and prometryn, *Pestic. Sci.,* 7, 41, 1976.
104. **Walker, A.,** Simulation of herbicide persistence in soil. III. Propyzamide in different soil types, *Pestic. Sci.,* 7, 59, 1976.
105. **Gingerich, L. L. and Zimdahl, R. L.,** Soil persistence of isopropalin and oryzalin, *Weed Sci.,* 24, 431, 1976.
106. **Walker, A.,** Persistence of pronamide in soil, *Pestic. Sci.,* 1, 237, 1970.
107. **Burschel, P.,** Untersuchungen uber das Verhalten von Simazin in Boden, *Weed Res.,* 1, 131, 1961.
108. **Walker, A. and Bond, W.,** Persistence of the herbicide AC 92, 553, *N*-(ethylpropyl)-2,6-dinitro-3,4-xylidine, in soils, *Pestic. Sci.,* 8, 359, 1977.
109. **Horowitz, M. and Herzlinger, G.,** Soil conditions affecting the dissipation of diuron, fluometuron and propham from the soil surface, *Weed Res.,* 14, 257, 1974.
110. **Weed, S. B. and Weber, J. B.,** Pesticide-organic matter interactions, in *Pesticides in Soil and Water,* Guenzi, W. D., Ed., Soil Science Society of America, Madison, Wisconsin, 1974, 39.
111. **Walker, A.,** A simulation model for prediction of herbicide persistence, *J. Environ. Qual.,* 3, 396, 1974.
112. **Walker, A. and Barnes, A.,** Simulation of herbicide persistence in soil: a revised computer model, *Pestic. Sci.,* 12, 123, 1981.

113. **Smith, A. E., Grover, R., Emmond, G. S., and Korven, H. C.,** Persistence and movement of atrazine, bromacil, monuron, and simazine in intermittently-filled irrigation ditches, *Can. J. Plant Sci.,* 55, 809, 1975.
114. **Thomas, R. G.,** Volatilization from soil, in *Handbook of Chemical Property Estimation Methods,* Lyman, W. J., Reehl, W. F., and Rosenblatt, D. H., Eds., McGraw-Hill, New York, 1982, chap. 16.
115. **Jury, W. A., Farmer, W. J., and Spencer, W. F.,** Behavior assessment model for trace organics in soil. II. Chemical classification and parameter sensitivity, *J. Environ. Qual.,* 13, 567, 1984.
116. **Hartley, G. S.,** Evaporation of pesticides, in *Pesticidal Formulations Research, Physical and Colloidal Chemical Aspects,* Advanced Chemistry Ser. 86, 1969, 115.
117. **Hamaker, J. W.,** Diffusion and volatilization, in *Organic Chemicals in the Soil Environment,* Goring, C. A. A. and Hamaker, J. W., Eds., Marcel Dekker, New York, 1972, 341.
118. **Mayer, R., Letey, J., and Farmer, W. J.,** Models for predicting volatilization of soil-incorporated pesticides, *Soil Sci. Soc. Am. Proc.,* 38, 563, 1974.
119. **Karickhoff, S. W., Brown, D. S., and Scott, T. A.,** Sorption of hydrophobic pollutants on natural sediments, *Water Res.,* 13, 241, 1979.
120. **Hassett, J. J., Means, J. C., Banwart, W. L., and Wood, S. G.,** Sorption Properties of Sediments and Energy-Related Pollutants, Final Report, *EPA-600/3-80-041,* 1980.
121. **Kenaga, E. E.,** Predicted bioconcentration factors and soil sorption coefficients of pesticides and other chemicals, *Ecotoxic. Environ. Safety,* 4, 26,1980.
122. **Beste, C. E., Ed.,** *Herbicide Handbook of the Weed Science Society of America,* 5th ed. Weed Science Society of America, Champaign, Illinois, 1983.
123. **Ferreira, G. A. L. and Seiber, J. N.,** Volatilizaiton and exudation losses of three *N*-methyl-carbamate insecticides applied systemically to rice, *J. Agric. Food Chem.,* 29, 93, 1981.
124. **Spencer, W. F. and Cliath, M. M.,** Vapor density of dieldrin, *Environ. Sci. Technol.,* 3, 670, 1969.
125. **Bowery, T. G.,** Heptachlor, in *Analytical Methods for Pesticides, Plant Growth Regulators, and Food Additives,* Vol. 2, Zweig, G., Ed., Academic Press, New York, 1964, 245.
126. **Spencer, W. F. and Cliath, M. M.,** Vapor density and apparent vapor pressure of lindane (γBHC), *J. Agric. Food Chem.,* 18, 529, 1970.
127. **Leonard, R. A., Bailey, G. W., and Swank, R. R., Jr.,** Transport, detoxification, fate and effects of pesticides in soil and water environments in *Land Application of Waste Materials,* Soil Conservation Society of America, Ankeny, Iowa, 1976, 48.
128. **Spencer, W. F., Shoup, T. D., Cliath, M. M., Farmer, W. J., and Haque, R.,** Vapor pressures and relative volatility of ethyl and methyl parathion, *J. Agric. Food Chem.,* 27, 273, 1979.
129. **Harper, L. A., White, A. W., Jr., Bruce, R. R., Thomas, A. W., and Leonard, R. A.,** Soil and microclimate effects on trifluralin volatilization, *J. Environ. Qual.,* 5, 236, 1976.
130. **Edwards, C. A.,** Nature and origins of pollution of aquatic systems by pesticides, in *Pesticides in Aquatic Environments,* Khan, M. A. Q., Ed., Plenum Press, New York, 1977, 11.
131. **Rao, P. S. C. and Davidson, J. M.,** Adsorption and movement in selected pesticides at high concentrations in soils, *Water Res.,* 13, 376, 1979.
132. **Schroy, J. M., Hileman, F. E., and Cheng, S. E.,** The uniqueness of dioxins? Physical/chemical characteristics, in 8th ASTM Aquatic Toxicology Symp., April 15 to 17, 1984, Fort Mitchell, Kentucky.
133. **McDowell, L. L., Willis, G. H., Murphree, C. E., Southwick, L. M., and Smith, S.,** Toxaphene and sediment yields in runoff from a Mississippi delta watershed, *J. Environ. Qual.,* 10, 120, 1981.
134. **Argauer, R. J., Mason, H. C., Corleyu, C., Higgens, A. H., Sauls, J. N., and Liljedahl, L. A.,** Drift of water-diluted and undiluted formulations and malathion and azinphosmethyl applied by airplane, *J. Econ. Entomol.,* 61, 1015, 1968.
135. **Baker, R. D. and Applegate, H. G.,** Effect of temperature and ultraviolet radiation on the persistence of methyl parathion and DDT in soils, *Argon. J.,* 62, 509, 1970.
136. **Seiber, J. M., Madden, S. C., McChesney, M. M., and Winterlin, W. N.,** Toxaphene dissipation from treated cotton field environments: component residual behaviour on leaves and in air, soil, and sediments determined by capillary gas chromatography, *J. Agric. Food Chem.,* 27, 284, 1979.
137. **Briggs, G. G.,** Degradation in soils, in *Proc., British Crop Protection Council Symp.: The Persistence of Insecticides and Herbicides,* Monogr. 17, British Crop Protection Council, London, 1976, 41.

Chapter 6

TRANSFORMATIONS IN SOIL

Allan E. Smith

TABLE OF CONTENTS

I. INTRODUCTION

Once present in the soil, herbicides undergo transformation by chemical and biological processes to products which themselves are transformed into progressively smaller molecules. Eventually complete mineralization of the herbicide into inorganic materials, such as carbon dioxide and ammonia, may occur. Thus, the persistence of a herbicide in the soil depends entirely on how quickly it is metabolized or transformed into degradation products.

The term metabolite strictly refers to changes induced by living cells, while transformation indicates changes which result from both chemical and biochemical processes. Since it is often difficult to distinguish between the two mechanisms in soil, and because hydrolytic reactions may occur as a result of both processes, the term transformation will be used in this chapter to denote change.

In the following pages, a brief description of the techniques used for the study of herbicide transformations will be given, then the chemical and biochemical processes by which herbicides can be degraded will be discussed. The kinetics of such reactions will briefly be described, and finally, the soil transformations of important herbicides from each major structural class will be examined, and known degradation products that have been isolated both under laboratory and field conditions enumerated.

As a result of soil degradation experiments conducted under laboratory conditions with (^{14}C)-labeled herbicides, a certain amount of radioactivity always remains in an unextracted form. Some of this nonextractable activity has been shown to result from such herbicide degradation products as carbon dioxide, small carbon chains, substituted anilines, and catechols, being incorporated directly into soil organic matter. However, studies of this nature will not be included in this chapter.

II. METHODS FOR STUDYING TRANSFORMATIONS IN SOILS

A. Laboratory Methods

Today the study of the transformation of herbicides in the soil is conducted almost exclusively with (^{14}C)-labeled compounds, which are incubated with moist, nonsterile soil for varying periods of time. The soils are then solvent extracted and the radioactive degradation products separated using thin-layer chromatographic, gas liquid chromatographic, and high pressure liquid chromatographic techniques. Identification of degradation products can be achieved by comparing Rf and retention time values with those of known compounds suspected of being transformation products. Final confirmation is usually provided by mass spectral analysis of the separated degradation products. Mass spectral analysis also provides information regarding the molecular weight and configuration of degradation products when their structures can not be foreseen.

Originally, degradation studies were conducted using nonradioactive herbicides and culture solutions, it being more practical to extract and analyze for degradation products in aqueous solution than in soil. Thus, a considerable amount of research was carried out on the metabolism of herbicides in liquid media using microbial cultures. These experiments frequently involved the use of high herbicidal concentrations, and although such studies were useful in demonstrating the biodegradability of herbicides, their relevance to the degradation in soils was limited. However, if a degradation product can be isolated and identified from culture solutions, then attempts can be made to isolate the same product from soils fortified with realistic amounts of the herbicide under test.

Several procedures have been adopted in recent years to study the transformation of herbicides in soil under laboratory conditions.[1] One of the earliest was that of soil perfusion. In this technique, an aerated solution of the herbicide is circulated through a soil column with samples of the solution being removed at intervals and analyzed for both herbicide remaining and for the presence of the degradation products. Unfortuantely, the experimental conditions are not those to be found under normal field conditions.

Simple soil treatments have been used by several laboratories to isolate transformation products of herbicides. Cartons or flasks of soil at known moisture and temperature conditions are incubated with radioactively labeled (or unlabeled) herbicides. At intervals, the soils are solvent extracted and analyses conducted to isolate and identify degradation products. These studies can provide much information, but are limited in that volatile transformation products and any (^{14}C)carbon dioxide liberated would escape from the lightly sealed systems and not be detected.

The development of the Bartha and Pramer flasks[2] enabled the production of (^{14}C)carbon dioxide from labeled herbicides incubated with soils to be measured. These flasks are essentially Erlenmeyer flasks with a side arm. Moistened soil, treated with the (^{14}C)-labeled herbicide, is placed in the flask which is closed with a stopper. Dilute sodium hydroxide solution is added to the side arm which is then sealed with a stopper pierced with a needle, the tip of which is beneath the top of the caustic solution. As degradation of the (^{14}C)-labeled herbicide occurs, evolved (^{14}C)carbon dioxide is trapped by the sodium hydroxide and converted into sodium (^{14}C)carbonate. Aliquots of this solution can be withdrawn at intervals, through the needle, and assayed for radioactivity. This apparatus has some limitations; volatile degradation products cannot be trapped, also, as the oxygen in the flask is used up and fresh air is drawn into the system through the needle, a buildup of nitrogen can occur. The latter difficulty can be overcome by attaching the flasks to a source of oxygen.

Flow-through systems probably have the most advantages, and least disadvantages of all the laboratory soil incubation procedures.[1] These systems allow measurement of volatility losses and evolution of (^{14}C)carbon dioxide. In principle, air, saturated with water vapor, is passed over the treated soil and the effluent gas then passes through gas absorption bottles containing solutions such as ethylene glycol, sulfuric acid, and sodium hydroxide to trap volatile products and carbon dioxide. The system can be used to monitor the degradation of (^{14}C)-labeled herbicides under aerobic and anaerobic conditions.

B. Field Methods

Laboratory studies provide much information regarding transformation products and are the only convenient method for trapping and identifying volatile transformation compounds. Hence, they are excellent for indicating what products may be formed from a particular herbicide under field conditions. Nevertheless, it is important to know the identity and concentration of degradation products actually formed in field soils following normal agricultural applications of a herbicide. A simple approach has been to sample field soils at regular intervals after treatment, and, following solvent extraction, use conventional analytical techniques, such as gas and liquid chromatography, to quantify degradation products using suspected transformation products as standards.

This approach will not detect volatile degradation products, products that are only formed transiently in the soil, or products that are formed in amounts that are insufficiently sensitive to detection by conventional chromatographic techniques. Some of these difficulties have been overcome by the use of (^{14}C)-labeled herbicides under field conditions. Because of cost and safety regulations, the areas to be treated must of necessity be small, but miniature field plots, or areas physically enclosed by metal pipes, have been treated with labeled herbicides.[3] The field soils can then be removed after various time intervals and analyzed for degradation products.

III. TRANSFORMATION REACTIONS IN SOILS

Today, about 150 synthetic chemicals are incorporated into a host of formulated herbicidal products. Once in the soil, what happens to these compounds? Fortunately, most herbicides are degraded in the environment and are transformed into progressively smaller molecules both by chemical and biochemical processes.[4-6] Chemical transformations are mainly hydrolytic in nature, being induced by factors such as soil moisture, soil temperature, and soil pH. Biochemical modifications of herbicides are achieved by enzyme systems of soil microflora so conditions that favor microbiological growth are likely to foster biochemical degradation of herbicides.

In view of the many thousands of synthetic molecules being introduced into the air, soil, and aquatic environments, it might be considered that the chemical and biochemical reactions that transform the parent molecules would be equally numerous. However, as a result of extensive research conducted over the last 40 years with hundreds of naturally occurring and synthetic molecules, both in soil and aqueous media, it has been demonstrated that certain molecular groupings are more prone to chemical and biochemical reactions than others. Thus, it has transpired that there are relatively few transformation reactions that are involved in the degradtion of chemicals.

The types of reactions to be discussed below are important for the transformation of herbicides. Of these 12 reaction types, 9 are biologically mediated, while 3 can occur as a result of both chemical and biochemical mechanisms. On the basis of these reactions, and given the structure of a herbicidal compound, it is now possible to predict transformation products likely to be formed in the soil.

A. Transformation Processes

Dealkylation — These reactions involve the biological removal of a methyl, or other alkyl, group from a nitrogen atom, and occur during the degradation of herbicides with dinitroaniline, triazine, and urea configurations.

$$R\text{-}N\binom{CH_3}{CH_3} \rightarrow R\text{-}N\binom{CH_3}{H} \rightarrow R\text{-}N\binom{H}{H}$$

Dealkoxylation — Herbicides, especially those with urea structures, may have a methoxy grouping biologically cleaved from a nitrogen atom.

$$R\text{-}N\binom{CH_3}{OCH_3} \rightarrow R\text{-}N\binom{CH_3}{H}$$

Decarboxylation — This reaction can occur as a result of both chemical and biochemical mechanisms and involves the elimination of carbon dioxide from herbicides possessing carboxyl groups.

$$R\text{-}COOH \rightarrow R\text{-}H + CO_2$$

Dehalogenation — With certain herbicides, notably those whose structures are based on substituted aliphatic acids, chlorine atoms are biologically replaced with hydrogen atoms.

$$R\text{-}Cl \rightarrow R\text{-}H$$

Ether cleavage — This important mechanism is chiefly responsible for the degradation of phenoxyalkanoate acid-type herbicides by biologically cleaving the bond between oxygen and carbon atoms.

$$R-O-CH_2-R_1 \rightarrow R-OH$$

Hydrolysis — The cleavage of molecules by water is one of the most important mechanisms involved in herbicide degradation. Such reactions can occur as a result of both chemical and biochemical processes and are responsible for the transformation of herbicides with amide, carbamate, ester, nitrile, thiocarbamate, triazine, and urea structures:

- Amides are converted to acids and amines.

$$R-CO-NH-R_1 \rightarrow R-COOH + R_1-NH_2$$

- Carbamates are converted to amines and alcohols.

$$R-NH-CO-O-R_1 \rightarrow R-NH_2 + R_1-OH$$

- Esters are converted to acids and alcohols.

$$R-CO-O-R_1 \rightarrow R-COOH + R_1-OH$$

- Nitriles are converted to amides and then to carboxylic acids.

$$R-CN \rightarrow R-CO-NH_2 \rightarrow R-COOH$$

- Thiocarbamates are converted to amines and thioalcohols.

$$\begin{matrix} R_1 \\ \\ R_2 \end{matrix}\!\!\!\!>\!N-CO-S-R \rightarrow \begin{matrix} R_1 \\ \\ R_2 \end{matrix}\!\!\!\!>\!NH + R-SH$$

- Triazines are converted to hydroxytriazines.

$$Tr-Cl \rightarrow Tr-OH$$

$$Tr-OCH_3 \rightarrow Tr-OH$$

$$Tr-SCH_3 \rightarrow Tr-OH$$

- Phenylureas are converted to anilines and amines.

$$Ar-NH-CO-N\!\!<\!\!\begin{matrix} R_1 \\ R_2 \end{matrix} \rightarrow Ar-NH_2 + HN\!\!<\!\!\begin{matrix} R_1 \\ R_2 \end{matrix}$$

Hydroxylation — These reactions involve the biological introduction of hydroxyl groups into aromatic or aliphatic functions.

$$Ar-H \rightarrow Ar-OH$$

$$Ar-COOH \rightarrow Ar-OH$$

$$R-CH_3 \rightarrow R-CH_2OH$$

Methylation — This involves the biological addition of a methyl group to an alcohol or phenol with formation of a methyl ether. Phenolic degradation products of 2,4-D and 2,4,5-T undergo such methylation.

$$R\text{–}OH \rightarrow R\text{–}OCH_3$$

$$Ar\text{–}OH \rightarrow Ar\text{–}OCH_3$$

Oxidation — Both chemical and biochemical mechanisms can oxidize susceptible groups.

$$R\text{–}CH_2OH \rightarrow R\text{–}CHO \rightarrow R\text{–}COOH$$

Beta-oxidation — Carbon atoms linked to aromatic ring systems can be biologically degraded by elimination of two carbon atoms from the chain. This mechanism is important in converting 2,4-DB to 2,4-D and MCPB to MCPA.

$$Ar\text{–}O\text{–}(CH_2)_3\text{–}COOH \rightarrow Ar\text{–}O\text{–}CH_2\text{–}COOH$$

Reduction — Nitro groups can be biologically reduced to amino groups, and this mechanism is responsible for transforming herbicides with dinitroaniline structures.

$$R\text{–}NO_2 \rightarrow R\text{–}NH_2$$

Ring fission — These complex reactions involve the biological rupture of aromatic ring systems and occur most frequently with phenoxyalkanoic acid herbicides.

B. Kinetics of Transformation Reactions

Clearly, the greater the number of transformation reactions that can occur on a herbicide molecule, and the greater the number of microbial species capable of carrying out the various transformations, the faster the hebicide will be degraded, and consequently, the shorter will be the persistence in the soil. It is also apparent that should more than one transformation reaction occur at the same time, then more than one degradtion product will be formed.

In the case of degradation products that can themselves undergo further transformation, their presence in the soil may be only transient before further modification occurs to still simpler molecules. In these events, certain transformation products may not be detected. On the other hand, some transformation products that are more persistent than the parent herbicide, build up in the soil. This is summarized in Figure 1, which compares the buildup of a transformation product B, formed from herbicide A in soil. It is assumed that breakdown occurs by first order kinetics and a half-life value of T has been assumed for the herbicide A and half-life values of $^1/_2$T, T, and 2T have been taken for the product B. It can be noted (Figure 1), that under these defined conditions, the amounts of B in the soil are less when the half-life of B is less than that of the herbicide and greater when its half-life is greater than that of A.

A more complex situation arises when herbicide A is transformed in the soil to product B, that itself is transformed to product C which, in turn, undergoes degradation. Amounts of B and C generated in the soil are summarized in Table 1; again, first-order kinetics are presumed to prevail and different half-life values have been assumed for the transformation of A to B, B to C and for C to other products.

From Figure 1 and Table 1, it can be noted that less than 50% of the applied herbicide should be present in the soil as transformation products B or C, indicating that amounts of such products will generally be much lower than the concentration of the initially applied herbicide. If the molecular weights of B and C are less than that of A, then the amounts of B and C in the soil, expressed on a ppm basis, will be proportionally lower.

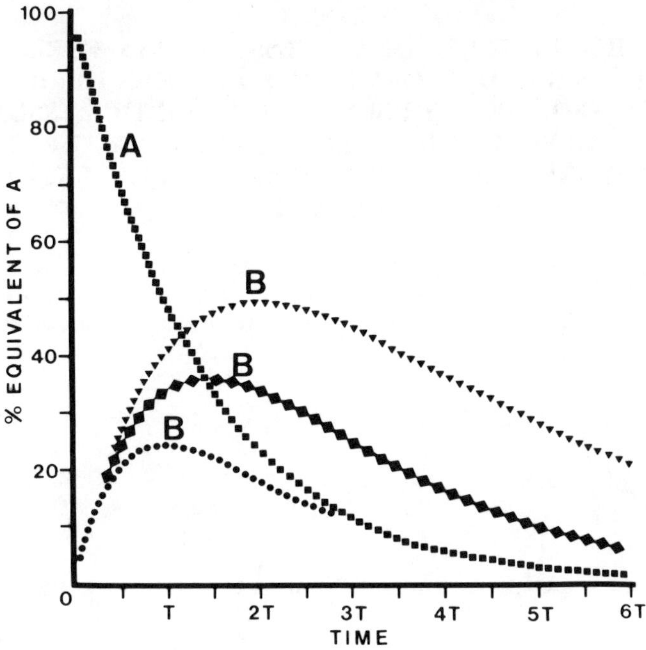

FIGURE 1.   Theoretical data depicting the breakdown of herbicide A (■) with formation of transformation product B, assuming first-order kinetics and a half-life of T for A and half-lives of $\frac{1}{2}$T (●), T (◆), and 2T (▼), respectively, for B.

IV. TRANSFORMATION REACTIONS BY HERBICIDE CLASSES

A. Aliphatic Acids

The fate, in soils, of the chlorinated aliphatic acids, of which dalapon (**1**) and TCA (**2**) are the most commonly encountered, have been extensively reviewed.[7,8]

$$CH_3 \cdot CCl_2 \cdot COOH \qquad\qquad CCl_3 \cdot COOH$$

$$\mathbf{1} \qquad\qquad\qquad\qquad \mathbf{2}$$

Under laboratory conditions both dalapon and TCA undergo transformation in moist, non-sterile soils by dehalogenation mechanisms. There is also evidence that the carbon atoms of the two herbicides can be released as carbon dioxide, or can be incorporated into cellular components of the degrading organisms. Thus, no discrete transformation products of either delapon or TCA have been isolated from soils under laboratory or field condtions.

B. Amides

The members of this heterogeneous class of compounds have a structure based on that of substituted amides. The structures of alachlor, butachlor, propachlor, propanil, and solan are compared in Table 2. The structure of CDAA (**3**) is that of an allylic amide, while a more complex molecular arrangement is shown by pronamide (**4**; R = –CO·NH·C(CH$_3$)$_2$·C≡CH).

Table 1
MAXIMUM RESIDUES OF SUCCESSIVE TRANSFORMATION PRODUCTS B AND C FORMED IN SOIL WHEN HERBICIDE A IS TRANSFORMED BY FIRST ORDER KINETICS ASSUMING DIFFERENT HALF-LIFE VALUES FOR THE TRANSFORMATION OF B TO C AND FOR C TO FURTHER PRODUCT D

Half-life $A \rightarrow B$ (days)	Half-life $B \rightarrow C$ (days)	Half-life $C \rightarrow D$ (days)	Max. residues of B (equiv. % of A) and days to reach max	Max. residues of C (equiv. % of A) and days to reach max
10	10	10	37 %, 14 days	27 %, 29 days
		20	37 %, 14 days	41 %, 36 days
		40	37 %, 14 days	55 %, 45 days
10	20	10	50 %, 20 days	20 %, 36 days
		20	50 %, 20 days	32 %, 46 days
		40	50 %, 20 days	46 %, 58 days
10	40	10	63 %, 27 days	14 %, 45 days
		20	63 %, 27 days	23 %, 58 days
		40	63 %, 27 days	35 %, 76 days
20	10	10	25 %, 20 days	20 %, 36 days
		20	25 %, 20 days	32 %, 46 days
		40	25 %, 20 days	46 %, 58 days
20	20	10	37 %, 29 days	16 %, 46 days
		20	37 %, 29 days	27 %, 58 days
		40	37 %, 29 days	41 %, 72 days
20	40	10	50 %, 40 days	12 %, 58 days
		20	50 %, 40 days	20 %, 72 days
		40	50 %, 40 days	32 %, 92 days
40	10	10	16 %, 27 days	14 %, 45 days
		20	16 %, 27 days	23 %, 58 days
		40	16 %, 27 days	36 %, 76 days
40	20	10	25 %, 40 days	12 %, 58 days
		20	25 %, 40 days	20 %, 72 days
		40	25 %, 40 days	32 %, 92 days
40	40	10	37 %, 58 days	9 %, 76 days
		20	37 %, 58 days	16 %, 92 days
		40	37 %, 58 days	27 %, 113 days

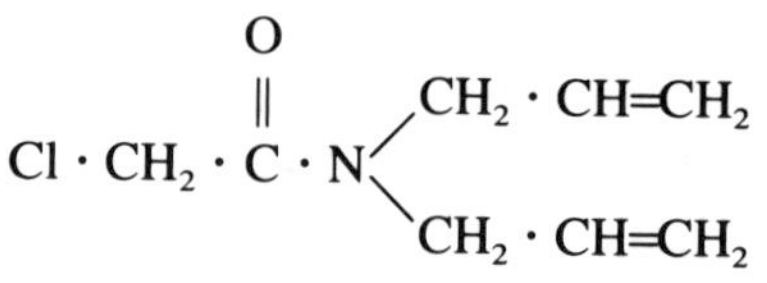

$$Cl \cdot CH_2 \cdot \overset{\overset{\textstyle O}{\|}}{C} \cdot N \Big\langle {}^{CH_2 \cdot CH=CH_2}_{CH_2 \cdot CH=CH_2}$$

3

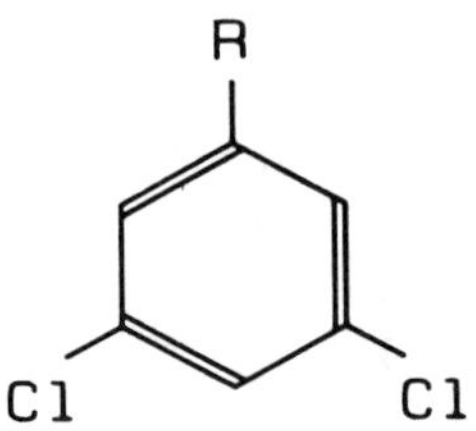

4

Table 2
COMMON NAMES AND STRUCTURES OF AMIDE HERBICIDES

Common name	R$_1$	R$_2$	R$_3$	R$_4$	R$_5$	R$_6$
Alachlor	–CH$_2$·O·CH$_3$	–CH$_2$·Cl	–CH$_2$·CH$_3$	–H	–H	–CH$_2$·CH$_3$
Butachlor	–CH$_2$·O(CH$_2$)$_3$·CH$_3$	–CH$_2$·Cl	–CH$_2$·CH$_3$	–H	–H	–CH$_2$·CH$_3$
Propachlor	–CH(CH$_3$)$_2$	–CH$_2$·Cl	–H	–H	–H	–H
Propanil	–H	–CH$_2$·CH$_3$	–H	–Cl	–Cl	–H
Solan	–H	–CH(CH$_3$)CH$_2$·CH$_2$·CH$_3$	–H	–Cl	–CH$_3$	–H

In the soil, these amide herbicides would be expected to undergo hydrolysis at the amide linkage with formation of an amine, or aniline, and a carboxylic acid. Dealkylation reactions are also possible, as is the cleavage of the ether bonds of alachlor and butachlor.

Although it has been reported[9] that (^{14}C)carbon dioxide is evolved from nonsterile soils treated with either carboxyl- or allyl-labeled (^{14}C)CDDA, no identifiable degradation products appear to have been isolated from CDAA-treated laboratory or field soils.

Soil enrichment studies conducted with high concentrations of propachlor, propanil, and solan have confirmed[10] that hydrolysis to substituted anilines can occur. It has been established[11-15] that when high concentrations of propanil undergo decomposition in soils, the resulting 3,4-dichloroaniline is converted by a self-coupling mechanism into 3,3′,4,4′-tetrachloroazobenzene. However, at normal herbicide treatment rates, the aniline has not been reported from field or laboratory soils, and it would seem that in these cases, the aniline is either degraded or becomes immobilized by being adsorbed to, or being incorporated into, soil organic matter.[16,17] Analysis of field soils treated with propanil have shown[18,19] that, in general, levels of tetrachloroazobenzene present in the soil were less than 0.05 ppm.

At low moisture levels and at temperatures between 38 and 46°C, alachlor can be transformed in soil to 2-chloro-2′,6′-diethylacetanilide.[20] Since the soil conditions would preclude biological mechanisms, this transformation must be chemical in origin.[20]

Laboratory studies with carboxyl-labeled (^{14}C)pronamide (**4**; R = –CO·NH·C(CH$_3$)$_2$·C≡CH) in nonsterile soil have indicated[21,22] that the herbicide can undergo cyclization to the methyleneoxazoline structure (**4**; R = **5**), with subsequent hydrolysis to the acetyl compound (**4**; R = –CO·NH·C(CH$_3$)$_2$·CO·CH$_3$).

5

Table 3
COMMON NAMES AND
STRUCTURES OF BENZOIC
ACID HERBICIDES

COOH structure: benzene ring with COOH at top, Cl and R_1 at positions flanking the top, R_2 and Cl at the bottom flanking positions.

Common name	R_1	R_2
Chloramben	–H	–NH$_2$
Dicamba	–OCH$_3$	–H
2,3,6-TBA	–Cl	–H
Tricamba	–OCH$_3$	–Cl

C. Benzoic Acids

The structures of herbicides classed as substituted benzoic acids are summarized in Table 3. Little is known regarding the transformation of chloramben, 2,3,6-TBA or tricamba in soils, though it has been reported[23] that (^{14}C)carbon dioxide can be evolved from nonsterile soils treated with carboxyl-labeled (^{14}C)chloramben, and that the chlorine atoms of 2,3,6-TBA can be released in the soil as chloride.[24]

Considerably more is known about the fate of dicamba in soils. This chemical, like the other benzoic acid herbicides, is applied as an amine salt. Once present in the soil, it has been established[25] that the dimethylamine salt of dicamba undergoes dissociation with the dimethylamine cation becoming strongly adsorbed to soil colloids. Thus, in the soil, dicamba will exist in the anionic state, as is the case for the phenoxyalkanoate herbicides.

Under laboratory conditions, in nonsterile soils, ether cleavage of the dicamba molecule occurs with formation of 3,6-dichlorosalicylic acid, which then undergoes further breakdown to unidentified degradation products.[26,27] Laboratory studies with (^{14}C)dicamba, have shown that both carboxyl- and ring-carbon atoms of dicamba can be released as (^{14}C)carbon dioxide[26,27] with loss of the carboxyl-carbon atom being considerably more rapid than ring fission mechanisms. Currently, it is not known whether it is the dicamba or the 3,6-dichlorosalicylic acid, or both, which undergoes decarboxylation and ring fission during soil incubation.

Investigations conducted using small field plots treated with ring-labeled (^{14}C)dicamba have indicated[28] that no (^{14}C)3,6-dichlorosalicylic acid was present in the soil 45 weeks after treatment with the (^{14}C)herbicide.

D. Benzonitriles

The structures of these herbicides, which are based on that of benzonitrile, are displayed in Table 4. Bromoxynil and ioxynil are applied as their alkali metal salts or as their octanoate esters. It has been noted[29] that once in moist soils, bromoxynil octanoate undergoes a very rapid hydrolysis to the phenolic form of bromoxynil. This hydrolysis is most likely chemical in origin, and similar to that responsible for the conversion of phenoxyalkanoate esters to their respective acid anions. A similar hydrolysis would be expected to convert ioxynil octanoate to the phenolic anion in moist soils.

All three compounds undergo hydrolysis in nonsterile soils with subsequent transformation of the cyano group to that of an amide.[30-35] In the cases of bromoxynil and ioxynil, further hydrolytic transformation of the amide to the acid can occur.[32,35,36]

Table 4
COMMON NAMES AND STRUCTURES OF
BENZONITRILE HERBICIDES

CN

R_5 R_1

R_4 R_2

R_3

Common name	R_1	R_2	R_3	R_4	R_5
Bromoxynil	–H	–Br	–OH	–Br	–H
Dichlobenil	–Cl	–H	–H	–H	–Cl
Ioxynil	–H	–I	–OH	–I	–H

Table 5
TRANSFORMATION PRODUCTS ISOLATED FROM LABORATORY AND
FIELD SOILS TREATED WITH BENZONITRILE HERBICIDES

Herbicide	Transformation mechanism	Transformation products	Field and/or lab conditions	Ref.
Dichlobenil	Nitrile hydrolysis	2,6-Dichlorobenzamide	Lab and field	30, 31, 34, 37
Bromoxynil	Nitrile and then amide hydrolysis	3,5-Dibromo-4-hydroxybenzamide	Lab	32
		3,5-Dibromo-4-hydroxybenzoic acid	Lab	32
Ioxynil	Nitrile and then amide hydrolysis	3,5-Diiodo-4-hydroxybenzamide	Lab	35
		3,5-Diiodo-4-hydroxybenzoic acid	Lab	35

Laboratory studies undertaken with ring-labeled (^{14}C)ioxynil have indicated that extensive ring fission occurs with evolution of (^{14}C)carbon dioxide.[36] It has also been reported[35] that iodide ions are liberated from laboratory soils treated with ioxynil.

Degradation products isolated from field and laboratory soils treated with benzonitrile herbicides are summarized in Table 5.

Buildup of 2,6-dichlorobenzamide, formed from dichlobenil, in field soils does not seem to occur, even though this product has been reported[33] to be very persistent in soils. Analysis of soil from an orchard receiving high and repeated applications of dichlobenil indicated a residue of approximately 1 ppm of the amide in the top 15 cm of the orchard soil.[37]

Table 6
COMMON NAMES AND STRUCTURES OF
CARBAMATE HERBICIDES

$$O$$
$$\|$$
$$NHCOR_1$$

Common name	R_1	R_2	R_3
Barban	$-CH_2 \cdot C \equiv C \cdot CH_2 \cdot Cl$	$-Cl$	$-H$
Chlorpropham	$-CH(CH_3)_2$	$-Cl$	$-H$
Propham	$-CH(CH_3)_2$	$-H$	$-H$
Swep	$-CH_3$	$-Cl$	Cl

E. Bipyridyls

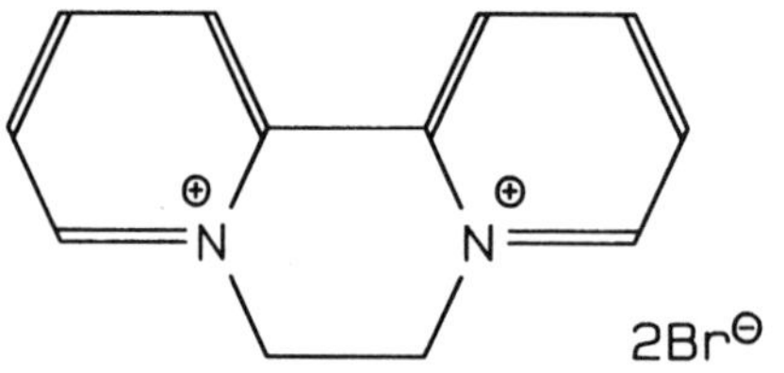

6

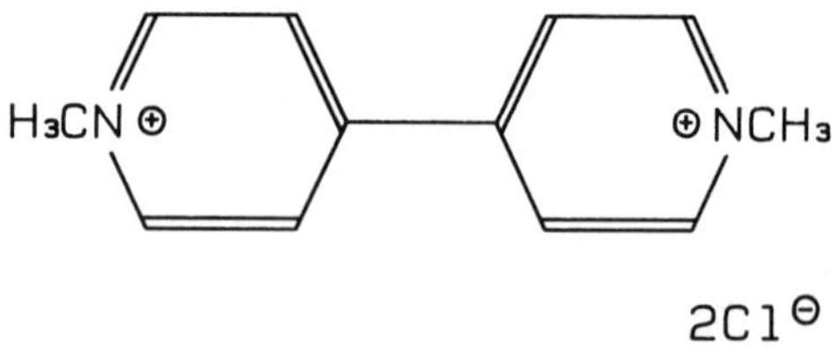

7

The quaternary bipyridylium herbicides diquat (**6**) and paraquat (**7**) are extensively used as crop desiccants and as rapidly acting nonselective herbicides. Both compounds are strongly adsorbed to soil by an ion exchange mechanism, and although there is evidence for their degradation in soils, no transformation products have been isolated and identified.[38]

F. Carbamates

The substituted carbamic acid esters are widely used for weed control, and the structures of barban, chlorpropham, propham, and swep are shown in Table 6. Another commonly encountered carbamate herbicide whose structure is slightly different is asulam (**8**; R = $-SO_2 \cdot NH \cdot CO \cdot OCH_3$).

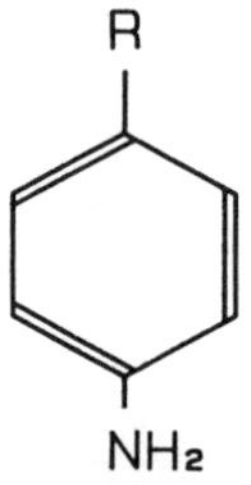

8

Table 7

COMMON NAMES AND STRUCTURES OF DINITROANILINE HERBICIDES

Common name	R_1	R_2	R_3	R_4
Benefin	$-CH_2 \cdot CH_3$	$-CH_2 \cdot CH_2 \cdot CH_2 \cdot CH_3$	$-H$	$-CF_3$
Butralin	$-H$	$-CH(CH_3)CH_2 \cdot CH_3$	$-H$	$-C(CH_3)_3$
Dinitramine	$-CH_2 \cdot CH_3$	$-CH_2 \cdot CH_3$	$-NH_2$	$-CF_3$
Fluchloralin	$-CH_2 \cdot CH_2 \cdot CH_3$	$-CH_2 \cdot CH \cdot_2 \cdot Cl$	$-H$	$-CF_3$
Isopropalin	$-CH_2 \cdot CH_2 \cdot CH_3$	$-CH_2 \cdot CH_2 \cdot CH_3$	$-H$	$-CH(CH_3)_2$
Oryzalin	$-CH_2 \cdot CH_2 \cdot CH_3$	$-CH_2 \cdot CH_2 \cdot CH_3$	$-H$	$-SO_2NH_2$
Profluralin	$-CH_2-\triangleleft$	$-CH_2 \cdot CH_2 \cdot CH_3$	$-H$	$-CF_3$
Trifluralin	$-CH_2 \cdot CH_2 \cdot CH_3$	$-CH_2 \cdot CH_2 \cdot CH_3$	$-H$	$-CF_3$

Since the review[15] on the fate of barban, chlorpropham, and swep in soils, little further information has been reported. As would be expected, the carbamate herbicides can undergo biological hydrolysis to their respective chlorinated anilines. Thus, studies with soil organisms have confirmed that barban and chlorpropham can be converted to 3-chloroaniline, while swep and propham are transformed into 3,4-dichloroaniline and aniline, respectively.[15] No degradation products have been isolated from field soils treated with carbamate herbicides. while swep and propham are transformed into 3,4-dichloroaniline and aniline, respectively.[15] No degradation products have been isolated from field soils treated with carbamate herbicides.

As in the case of the amide herbicides, it was reported that subsequent transformation of substituted anilines, formed from carbamate structures, to chlorinated azobenzenes was possible.[39,40] However, these studies were all conducted with unrealistically high aniline concentrations, and, as noted earlier, the amounts of anilines formed under field conditions from normal herbicidal applications of carbamates would be lower than those concentrations at which self-coupling aniline products are reported to occur.

Laboratory studies with asulam have revealed[41] that small amounts of sulfanilamide (**8**; $R = -SO_2 \cdot NH_2$) can be recovered from asulam-treated soils as a result of hydrolytic transformation. In moist, nonsterile soils, sulfanilamide is rapidly degraded to unidentifiable products.[42]

G. Dinitroanilines

This very important class of soil-applied herbicides have configurations based on those of substituted 2,6-dinitroanilines, and the structures of the most important members of the series are displayed in Table 7.

In recent years, the fate of these compounds has been extensively studied under laboratory and field conditions, and the findings reviewed.[43-50] The degradation of the dinitroaniline herbicides is complex, being different under aerobic and anaerobic conditions. In this chapter, only transformations encountered under aerobic situations will be discussed.

Transformation mechanisms met during the degradation of these compounds include dealkylation, reduction of nitro groups to amino groups, and cyclization to benzimadazole derivatives, which themselves can undergo dealkylation transformations. It has also been observed[44] that traces of dimeric compounds can be formed from transformation products of trifluralin.

Structures of transformation products isolated from moist, nonsterile soils treated with ring-labeled (^{14}C)dinitroaniline herbicides under laboratory and greenhouse conditions are summarized in Table 8. Amounts of the various products recovered seem always to be low and usually account for less than 5% of the applied herbicide.

The elegant and careful research conducted by Golab and associates[43-45] with ring-labeled (^{14}C)dinitroaniline herbicides under field conditions has led to the isolation of 9 transformation products from oryzalin, 12 from isopropalin, and 28 from trifluralin. None of these compounds accounted for more than 4% of the initial herbicidal treatment, and many were present in the soil in trace amounts only.

The major degradation products recovered from field soils treated with isopropalin, oryzalin, and trifluralin are compared in Table 9, and result from dealkylation and cyclization transformations.

Both laboratory and field studies indicate that the transformation products derived from dinitroaniline herbicides can undergo further modification in the soil to products that become strongly bound to, or incorporated into, soil organic matter.[44]

H. Diphenylethers

Bifenox (**13**; R_1 = –NO$_2$, R_2 = R_3 = –Cl, R_4 = –CO$_2$CH$_3$), fluorodifen (**13**; R_1 = R_2 = –NO$_2$, R_3 = –CF$_3$, R_4 = –H), and nitrofen (**13**; R_1 = –NO$_2$, R_2 = R_3 = –Cl, R_4 = –H) are substituted diphenylether herbicides. Although originally developed for the control of broadleaved weeds in rice paddy fields, they can also be used as preemergence treatments for controlling weeds in a vareity of crops.

13

Under flooded conditions, reduction of the nitro group of nitrofen can occur in soil with formation of the amino derivative[53] (**13**; R_1 = –NH$_2$, R_2 = R_3 = –Cl, R_4 = –H).

Greenhouse and laboratory studies.[54] with soils treated with (^{14}C)bifenox have resulted in the identification of small amounts of the following transformation products:

13; R_1 = –NO$_2$, R_2 = R_3 = –Cl, R_4 = –CO$_2$H

13; R_1 = –NH$_2$, R_2 = R_3 = –Cl, R_4 = –CO$_2$H

This would indicate that in soils, the transformation reactions of hydrolysis and reduction play a part in degrading the herbicides of this class. No transformation products have been isolated from field soils treated with these chemicals.

I. Phenols

The herbicides dinoseb (**14**; R = –CH(CH$_3$)CH$_2$·CH$_3$) and DNOC (**14**; R = –CH$_3$) have structures based on 6-substituted 2,4-dinitrophenol (**14**). These compounds were developed

Table 8
TRANSFORMATION PRODUCTS ISOLATED FROM LABORATORY AND GREENHOUSE SOILS TREATED WITH DINITROANILINE HERBICIDES

Herbicide	Transformation mechanism	Structure	R_1	R_2	R_3	R_4	R_5	Ref.
Benefin	Dealkylation	9	$-NH_2$	$-NH_2$	$-H$	$-CF_3$	$-NO_2$	51
	Dealk. + redn.	9	$-NH_2$	$-NH_2$	$-H$	$-CF_3$	$-NH_2$	51
		9	$-NH(C_4H_9)$	$-NH_2$	$-H$	$-CF_3$	$-NH_2$	51
	Dealkylation + hydroxylation	9	$-OH$	$-NO_2$	$-H$	$-CF_3$	$-NO_2$	51
Butralin	Dealkylation + loss of nitro	9	$-NH_2$	$-NO_2$	$-H$	$-C(CH_3)_3$	$-NO_2$	48, 49
		9	$-NH \cdot CH(CH_3)C_2H_5$	$-H$	$-H$	$-C(CH_3)_3$	$-NO_2$	48
Dinitramine	Dealkylation	9	$-NH(C_2H_5)$	$-NO_2$	$-NH_2$	$-CF_3$	$-NO_2$	48, 52
		9	$-NH_2$	$-NO_2$	$-NH_2$	$-CF_3$	$-NO_2$	48
	Cyclization	10	$-C_2H_5$	$-CH_3$	$-CF_3$	$-NH_2$	$-NO_2$	48,52
	Cycl. + dealk.	10	$-H$	$-CH_3$	$-CF_3$	$-NH_2$	$-NO_2$	48,52
Fluchloralin	Dealkylation	9	$-NH \cdot CH_2 \cdot CH_2 \cdot Cl$	$-NO_2$	$-H$	$-CF_3$	$-NO_2$	48
		9	$-NH(C_3H_7)$	$-NO_2$	$-H$	$-CF_3$	$-NO_2$	48
		9	$-NH_2$	$-NO_2$	$-H$	$-CF_3$	$-NO_2$	48
	Cycl. + dealk.	10	$-H$	$-C_2H_5$	$-CF_3$	$-H$	$-NO_2$	48
Profluralin	Dealkylation	9	$-NH(CH_2-\triangleleft)$	$-NO_2$	$-H$	$-CF_3$	$-NO_2$	48
		9	$-NH(C_3H_7)$	$-NO_2$	$-H$	$-CF_3$	$-NO_2$	48
Trifluralin	Dealkylation	9	$-NH(C_3H_7)$	$-NO_2$	$-H$	$-CF_3$	$-NO_2$	48,50
		9	$-NH_2$	$-NO_2$	$-H$	$-CF_3$	$-NO_2$	48,50
	Cyclization	10	$-C_3H_7$	$-C_2H_5$	$-CF_3$	$-H$	$-NO_2$	48,50
	Cycl. + dealk.	10	$-H$	$-C_2H_5$	$-CF_3$	$-H$	$-NO_2$	48,50

Table 9
TRANSFORMATION PRODUCTS ISOLATED FROM FIELD SOILS TREATED WITH DINITROANILINE HERBICIDES

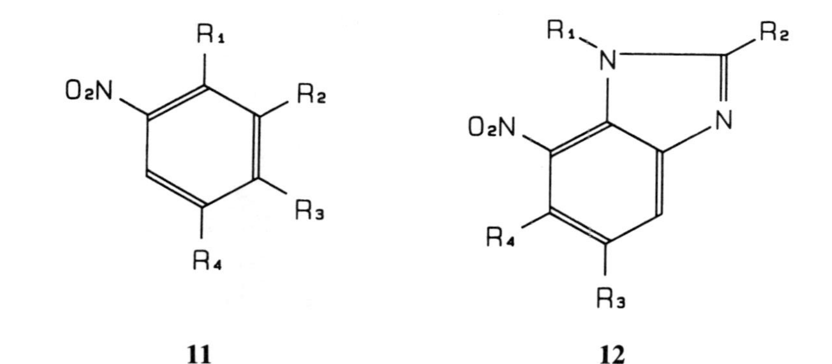

11   12

Herbicide	Transformation mechanism	Structure	R_1	R_2	R_3	R_4	Ref.
Isopropalin	Dealkylation	11	$-NH(C_3H_7)$	$-NO_2$	$-H$	$-CH(CH_3)_2$	43
		11	$-NH_2$	$-NO_2$	$-H$	$-CH(CH_3)_2$	43
	Hydroxylation	11	$-N(C_3H_7)_2$	$-NO_2$	$-H$	$-C \cdot OH(CH_3)_2$	43
	Cyclization	12	$-C_3H_7$	$-C_2H_5$	$-CH(CH_3)_2$	$-H$	43
	Cycl. + dealk.	12	$-H$	$-C_2H_5$	$-CH(CH_3)_2$	$-H$	43
Oryzalin	Dealkylation	11	$-NH(C_3H_7)$	$-NO_2$	$-H$	$-SO_2NH_2$	45
	Cyclization	12	$-C_3H_7$	$-C_2H_5$	$-SO_2NH_2$	$-H$	45
	Cycl. + dealk	12	$-H$	$-C_2H_5$	$-SO_2NH_2$	$-H$	45
Trifluralin	Dealkylation	11	$-NH(C_3H_7)$	$-NO_2$	$-H$	$-CF_3$	44
	Dealk. + hydrox.	11	$-OH$	$-NO_2$	$-H$	$-CF_3$	44
	Cyclization	12	$-C_3H_7$	$-C_2H_5$	$-CF_3$	$-H$	44
	Cycl. + dealk	12	$-H$	$-C_2H_5$	$-CF_3$	$-H$	44
		12	$-C_3H_7$	$-H$	$-CF_3$	$-H$	44

Table 10
COMMON NAMES AND STRUCTURES OF
PHENOXYALKANOIC ACID HERBICIDES

Common name	R_1	R_2	R_3	R_4
MCPA	$-CH_2 \cdot CO_2H$	$-CH_3$	$-Cl$	$-H$
2,4-D	$-CH_2 \cdot CO_2H$	$-Cl$	$-Cl$	$-H$
2,4,5-T	$-CH_2 \cdot CO_2H$	$-Cl$	$-Cl$	$-Cl$
Dichlorprop	$-CH(CH_3)CO_2H$	$-Cl$	$-Cl$	$-H$
Mecoprop	$-CH(CH_3)CO_2H$	$-CH_3$	$-Cl$	$-H$
MCPB	$-CH_2 \cdot CH_2 \cdot CH_2 \cdot CO_2H$	$-CH_3$	$-Cl$	$-H$
2,4-DB	$-CH_2 \cdot CH_2 \cdot CH_2 \cdot CO_2H$	$-Cl$	$-Cl$	$-H$

during the 1930s and early 1940s, and are thus among the oldest of organic herbicides still in existence. Perhaps for this reason, little research on their fate in soil has been reported since the review of Kaufman.[55]

14

Although it would appear from laboratory studies[55] with microbial cultures that reduction of a nitro group to that of an amino can occur, as can oxidative elimination of a nitro group with formation of a dihydric phenol, no transformation products have been isolated from treated soils under field or laboratory conditions.

J. Phenoxyalkanoic Acids

The structures of these important herbicides, which are based on chlorine-substituted phenoxyacetic, phenoxypropionic, and phenoxybutyric acids, are shown in Table 10. The phenoxyalkanoic acid herbicides are formulated as substituted amines, esters, and alkali metal salts.

Once present in the soil, it has been established that the ester formulations undergo rapid hydrolysis to the phenoxyalkanoate anion. Thus, the *iso*-propyl, *n*-butyl, and *iso*-octyl esters of 2,4-D are rapidly converted to the 2,4-D acid anion in soils with moisture contents above that of the wilting point.[56-58] Similarly, the *iso*-propyl, *n*-butyl, and *iso*-octyl esters of both MCPA and 2,4,5-T are hydrolyzed in soils to their corresponding anions.[59-61] Esters of the

Table 11

**TRANSFORMATION PRODUCTS ISOLATED FROM LABORATORY
AND FIELD SOILS TREATED WITH PHENOXYALKANOIC ACID
HERBICIDES**

Herbicide	Transformation mechanism	Transformation products	Field and/or lab conditions	Ref.
MCPA	Ether cleavage	2-Methyl-4-chlorophenol	Field	72
2,4-D	Ether cleavage	2,4-Dichlorophenol	Lab	65
	Ether cleavage + methylation	2,4-Dichloroanisole	Lab	65
2,4,5-T	Ether cleavage	2,4,5-Trichlorophenol	Lab	67
	Ether cleavage + methylation	2,4,5-Trichloroanisole	Lab	67
Mecoprop	Ether cleavage	2-Methyl-4-chlorophenol	Lab	66
MCPB	*Beta*-oxidation	MCPA	Lab	69
2,4-DB	*Beta*-oxidation	2,4-D	Lab	68

phenoxypropionic and phenoxybutyric acids behave analogously with the *iso*-octyl ester of dichlorprop and the *n*-butyl ester of 2,4-DB undergoing rapid hydrolysis in moist soils to their corresponding phenoxyalkanoic anions.[60] It has been considered[58,60,61] that the rapidity of hydrolysis in moist soils tended to preclude biological mechanisms, and that the ester hydrolysis was predominately due to a soil-catalyzed reaction. Studies[62] have also shown that in moist soils, the dimethylamine salt of 2,4-D can undergo dissociation with the amine cation becoming strongly adsorbed to soil colloids.

From these experiments it can be concluded that any phenoxyalkanoic herbicide, whether applied as ester or dimethylamine salt formulations, will be transformed to the respective phenoxyalkanoic anion in soil. These anions would then reassociate with inorganic cations present in the soil before undergoing biological degradation.

In moist, nonsterile soils, it has been observed that phenoxyalkanoic acids are biologically transformed with cleavage occurring between the oxygen atom and the carbon atom at the 2-position of the side chain.[63-67] The phenols thus formed, can then undergo biological methylation to the correspondingly substituted anisoles.[65,67] *Beta*-oxidation also appears to be an important mechanism whereby phenoxybutyric acids are converted in the soil to phenoxyacetic acids.[68,69]

Laboratory studies with ring-labeled (^{14}C)2,4-D,[65,67,70,71] ring-labeled (^{14}C)2,4,5-T,[67] and ring-labeled (^{14}C)mecoprop[66] have confirmed that the aromatic ring system of phenoxyalkanoic acid herbicides can be degraded in soil to (^{14}C)carbon dioxide.

Degradation products isolated from soils treated under laboratory and field conditions with phenoxyalkanoic acids are summarized in Table 11. 2-Methyl-4-chlorophenol has been identified[72] in field soils treated with MCPA.

Because of the volatile nature of chlorinated phenols and anisoles, their presence in field soils can only be of short duration.

K. Thiocarbamates

The structures of the most important compounds belonging to this class of soil-incorporated herbicides are compared in Table 12. Little is known regarding the fate of these compounds in soil. Laboratory studies conducted with (^{14}C-ethyl)EPTC,[73] (^{14}C-carbonyl)diallate,[74-76] and

Table 12
COMMON NAMES AND STRUCTURES OF THIOCARBAMATE HERBICIDES

$$R_1 \cdot S \cdot \overset{\displaystyle O}{\overset{\displaystyle \|}{C}} \cdot N \overset{R_2}{\underset{R_3}{<}}$$

Common name	R_1	R_2	R_3
Diallate	–CH$_2$·CCl=CHCl	–CH(CH$_3$)$_2$	–CH(CH$_3$)$_2$
EPTC	–CH$_2$·CH$_3$	–CH$_2$·CH$_2$·CH$_3$	–CH$_2$·CH$_2$·CH$_3$
Pebulate	–CH$_2$·CH$_2$·CH$_3$	–CH$_2$·CH$_3$	–CH$_2$·CH$_2$·CH$_2$·CH$_3$
Triallate	–CH$_2$·CCl=CCl$_2$	–CH(CH$_3$)$_2$	–CH(CH$_3$)$_2$
Vernolate	–CH$_2$·CH$_2$·CH$_3$	–CH$_2$·CH$_2$·CH$_3$	–CH$_2$·CH$_2$·CH$_3$

Table 13
COMMON NAMES AND STRUCTURES OF
S-TRIAZINE HERBICIDES

Common name	R_1	R_2	R_3
Atrazine	–Cl	–CH$_2$·CH$_3$	–CH(CH$_3$)$_2$
Cyanazine	–Cl	–C(CH$_3$)$_2$CN	–CH$_2$·CH$_3$
Propazine	–Cl	–CH(CH$_3$)$_2$	–CH(CH$_3$)$_2$
Simazine	–Cl	–CH$_2$·CH$_3$	–CH$_2$·CH$_3$
Ametryn	–SCH$_3$	–CH(CH$_3$)$_2$	–CH$_2$·CH$_3$
Prometryn	–SCH$_3$	–CH(CH$_3$)$_2$	–CH(CH$_3$)$_2$
Prometon	–OCH$_3$	–CH(CH$_3$)$_2$	–CH(CH$_3$)$_2$

(allyl-2-^{14}C)triallate[75,76] have indicated (^{14}C)carbon dioxide to be the only degradation product identified. Thus, although it would be expected that transformation reactions should occur through dealkylation and hydrolytic mechanisms, no transformation products have been isolated from treated soils, suggesting that any transformation products formed may themselves be degraded in the soil, or volatilize from the soils more rapidly than the parent herbicides.

L. Triazines

With few exceptions, the members of this very important class of herbicides are based on a s-triazine structure containing two substituted amino groups, while the third ring carbon atom possesses a chloro-, methoxy-, or thiomethyl grouping. The structures of the most commonly used s-triazine herbicides are displayed in Table 13. Two important triazines whose molecular arrangements are different are hexazinone (**15**; R_1 = –N(CH$_3$)$_2$, R_2 = –CH$_3$, R_3 = –H) and metribuzin (**16**; R = –NH$_2$).

Table 14
TRANSFORMATION PRODUCTS ISOLATED FROM LABORATORY SOILS TREATED WITH TRIAZINE HERBICIDES

Herbicide	Transformation mechanism	R_1	R_2	R_3	Ref.
Atrazine	Hydrolysis	–OH	–CH$_2$·CH$_3$	–CH(CH$_3$)$_2$	77,81,82
	Dealkylation	–Cl	–CH$_2$·CH$_3$	–H	82
		–Cl	–H	–CH(CH$_3$)$_2$	82
		–Cl	–H	–H	82
Cyanazine	Hydrolysis	–Cl	–C(CH$_3$)$_2$CONH$_2$	–CH$_2$·CH$_3$	82,83
		–Cl	–C(CH$_3$)$_2$CO$_2$H	–CH$_2$·CH$_3$	82,83
		–OH	–C(CH$_3$)$_2$CO$_2$H	–CH$_2$·CH$_3$	82,83
	Dealkylation	–Cl	–H	–CH$_2$·CH$_3$	82,83
Propazine	Hydrolysis	–OH	–CH(CH$_3$)$_2$	–CH(CH$_3$)$_2$	77,81
Prometryn	Hydrolysis	–OH	–CH(CH$_3$)$_2$	–CH(CH$_3$)$_2$	84,85
	Dealkylation	–SCH$_3$	–CH(CH$_3$)$_2$	–H	85
Simazine	Hydrolysis	–OH	–CH$_2$·CH$_3$	–CH$_2$·CH$_3$	77,81,82
	Dealkylation	–Cl	–CH$_2$·CH$_3$	–H	82
		–Cl	–H	–H	82

15 **16**

The transformation of *s*-triazine herbicides in the soil has been reviewed,[77,78] and mechanisms resulting in triazine degradation are those of dealkylation and hydrolysis to hydroxytriazines. Ring cleavage does not seem to be a major degradation pathway.

The main transformation products isolated from triazine-treated soils under laboratory and field conditions are enumerated in Tables 14 and 15, respectively.

The fate of (^{14}C)hexazinone under laboratory and field conditions has been studied[6,79] and the major soil transformation products isolated and identified. The structures of these hexazinone degradation products are:

Table 15

TRANSFORMATION PRODUCTS ISOLATED FROM FIELD SOILS TREATED WITH TRIAZINE HERBICIDES

Herbicide	Transformation mechanism	R_1	R_2	R_3	Ref.
Atrazine	Dealkylation	–Cl	–H	–CH(CH$_3$)$_2$	86-91
		–Cl	–CH$_2$·CH$_3$	–H	89,91
	Hydrolysis	–OH	–CH$_2$·CH$_3$	–CH(CH$_3$)$_2$	87-90, 92
	Hydrol. + dealk.	–OH	–H	–CH(CH$_3$)$_2$	87, 88, 90
		–OH	–CH$_2$·CH$_3$	–H	87, 88
Cyanazine	Dealkylation	–Cl	–C(CH$_3$)$_2$CN	–H	93
		–Cl	–H	–CH$_2$·CH$_3$	81
	Hydrolysis	–Cl	–C(CH$_3$)$_2$CONH$_2$	–CH$_2$·CH$_3$	89-91, 93, 94
	Hydrol. + dealk.	–Cl	–C(CH$_3$)$_2$CONH$_2$	–H	93
Prometryn	Dealkylation	–SCH$_3$	–H	–CH(CH$_3$)$_2$	84
	Hydrolysis	–OH	–CH(CH$_3$)$_2$	–CH(CH$_3$)$_2$	84
Simazine	Hydrolysis	–OH	–CH$_2$·CH$_3$	–CH$_2$·CH$_3$	95

15; R_1 = –NHCH$_3$, R_2 = –CH$_3$, R_3 = –OH

15; R_1 = –NHCH$_3$, R_2 = –CH$_3$, R_3 = –H

15; R_1 = –NHCH$_3$, R_2 = –H, R_3 = –H

15; R_1 = –N(CH$_3$)$_2$, R_2 = –CH$_3$, R_3 = –OH

thus indicating that both dealkylation and hydrolytic transformations are involved in hexazinone breakdown.

Under field conditions, two major transformation products have been isolated from metribuzin-treated soils. These are (**17**; R = –NH$_2$) and (**17**; R = –H).[80] Traces of a third compound (**16** = –H) have also been detected in treated field soils.[80]

17

18

M. Uracils

The most commonly encountered herbicides belonging to the substituted uracils are bromacil (**18**; R_1 = –CH(CH$_3$)CH$_2$·CH$_3$, R_2 = –Br, R_3 = –CH$_3$) and terbacil (**18**; R_1 = –C(CH$_3$)$_3$, R_2 = –Cl, R_3 = –CH$_3$).

Table 16
COMMON NAMES AND STRUCTURES OF
PHENYLUREA HERBICIDES

Common name	R_1	R_2	R_3	R_4
Chlorbromuron	–OCH$_3$	–CH$_3$	–Br	–Cl
Chlortoluron	–CH$_3$	–CH$_3$	–CH$_3$	–Cl
Diuron	–CH$_3$	–CH$_3$	–Cl	–Cl
Fenuron	–CH$_3$	–CH$_3$	–H	–H
Fluometuron	–CH$_3$	–CH$_3$	–H	–CF$_3$
Isoproturon	–CH$_3$	–CH$_3$	–CH(CH$_3$)$_2$	–H
Linuron	–OCH$_3$	–CH$_3$	–Cl	–Cl
Monolinuron	–OCH$_3$	–CH$_3$	–Cl	–H
Monuron	–CH$_3$	–CH$_3$	–Cl	–H
Neburon	–CH$_3$	–(CH$_2$)$_3$CH$_3$	–Cl	–Cl

In moist, nonsterile soils, under laboratory conditions, it has been noted[96,97] that both
(2-^{14}C)bromacil and (2-^{14}C)terbacil undergo ring fission with slow release of (^{14}C)carbon
dioxide.

Field studies conducted with (^{14}C)bromacil and (^{14}C)terbacil have indicated that over 90%
of the extractable radioactivity remaining in the soil after 1 year was the parent compound,[96]
suggesting that any transformation products were transitory in the soil. In the case of
(2-^{14}C)bromacil-treated field soils, trace amounts of three transformation products were
isolated and tentatively identified as

$$\textbf{18;} \quad R_1 = -CH(CH_3)CH_2 \cdot CH_3, \quad R_2 = -Br, \quad R_3 = -CH_2OH$$

$$\textbf{18;} \quad R_1 = -CH(CH_3)CH(OH)CH_3, \quad R_2 = -Br, \quad R_3 = -CH_3$$

$$\textbf{18;} \quad R_1 = -CH(CH_3)CH_2 \cdot CH_2OH, \quad R_2 = -Br, \quad R_3 = -CH_3$$

indicating that bromacil can be transformed in the soil by hydroxylation of the side chain
alkyl groups. No degradation compounds were isolated from any of the (2-^{14}C)terbacil-
treated field plots.[96]

N. Ureas

The substituted urea herbicides are derivatives of urea (H$_2$N·CO·NH$_2$), and the first compounds to be synthesized contained an amino moiety fully substituted with alkyl or alkoxy
groups, while the second amino function contained a mono- or dihalogenated phenyl group.
Currently, there are about 20 such phenylurea compounds registered as herbicides, and the
structures of ten of the most commonly encountred compounds are displayed in Table 16.
In recent years, there have been a few heterocyclic urea herbicides introduced, but since
these are not extensively used, their transformation in soils will not be discussed.

Table 17
TRANSFORMATION PRODUCTS ISOLATED FROM
LABORATORY SOILS TREATED WITH UREA
HERBICIDES

Herbicide	Transformation mechanism	R_1	R_2	R_3	Ref.
Chlorbromuron	Dealkylation	$-NH \cdot CO \cdot NH \cdot OCH_3$	$-Br$	$-Cl$	100
	Dealkoxylation	$-NH \cdot CO \cdot NH \cdot CH_3$	$-Br$	$-Cl$	100
Chlortoluron	Dealkylation	$-NH \cdot CO \cdot NH \cdot CH_3$	$-CH_3$	$-Cl$	101
Fluometuron	Dealkylation	$-NH \cdot CO \cdot NH \cdot CH_3$	$-H$	$-CF_3$	102
		$-NH \cdot CO \cdot NH_2$	$-H$	$-CF_3$	102
	Hydrolysis	$-NH_2$	$-H$	$-CF_3$	102
Isoproturon	Dealkylation	$-NH \cdot CO \cdot NH \cdot CH_3$	$-CH(CH_3)_2$	$-H$	103
		$-NH \cdot CO \cdot NH_2$	$-CH(CH_3)_2$	$-H$	103
	Hydrolysis	$-NH_2$	$-CH(CH_3)_2$	$-H$	103
	Dealkylation + hydroxylation	$-NH \cdot CO \cdot NH \cdot CH_3$	$-COH(CH_3)_2$	$-H$	103
		$-NH \cdot CO \cdot NH_2$	$-COH(CH_3)_2$	$-H$	103
	Hydrolysis + hydroxylation	$-NH_2$	$-COH(CH_3)_2$	$-H$	103
Linuron	Dealkylation	$-NH \cdot CO \cdot NH \cdot OCH_3$	$-Cl$	$-Cl$	99
	Dealkoxylation	$-NH \cdot CO \cdot NH \cdot CH_3$	$-Cl$	$-Cl$	99
	Dealkylation + dealkoxylation	$-NH \cdot CO \cdot NH_2$	$-Cl$	$-Cl$	99
	Hydrolysis	$-NH_2$	$-Cl$	$-Cl$	99
Monolinuron	Dealkylation	$-NH \cdot CO \cdot NH \cdot OCH_3$	$-Cl$	$-H$	99
	Dealkoxylation	$-NH \cdot CO \cdot NH \cdot CH_3$	$-Cl$	$-H$	99
	Dealkylation + dealkoxylation	$-NH \cdot CO \cdot NH_2$	$-Cl$	$-H$	99
	Hydrolysis	$-NH_2$	$-Cl$	$-H$	99

The transformation of phenylurea herbicides in soils has been reviewed.[98,99] In moist, nonsterile soils, dealkylation and dealkoxylation mechanisms occur to transform the herbicides into mono- and unsubstituted phenylureas. Hydrolysis can also occur at the amide linkage with formation of substituted anilines. In general, the amounts of anilines recovered from soils following treatment with phenylureas are very low, and their conversion into chlorinated azobenzenes has not been reported. As noted earlier, at low soil concentrations, chlorinated anilines are considered to be incorporated into soil organic matter.

The main transformation products isolated from phenylurea-treated soils, both under laboratory and field conditions, are listed in Tables 17 and 18.

O. Miscellaneous

In this category will be placed eight commonly used herbicides whose structures preclude their being placed in the previously discussed classifications. These herbicides include ami-

Table 18

**TRANSFORMATION PRODUCTS ISOLATED FROM FIELD SOILS
TREATED WITH UREA HERBICIDES**

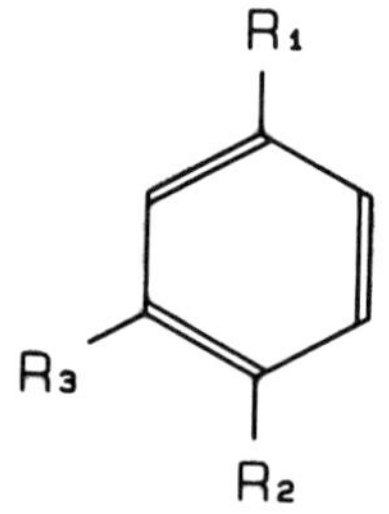

Herbicide	Transformation mechanism	R_1	R_2	R_3	Ref.
Chlortoluron	Dealkylation	$-NH \cdot CO \cdot NH \cdot CH_3$	$-CH_3$	$-Cl$	104
		$-NH \cdot CO \cdot NH_2$	$-CH_3$	$-Cl$	104
	Oxidation	$-NH \cdot CO \cdot N(CH_3)_2$	$-CO_2H$	$-Cl$	104
	Oxidation + dealkylation	$-NH \cdot CO \cdot NH \cdot CH_3$	$-CO_2H$	$-Cl$	104
		$-NH \cdot CO \cdot NH_2$	CO_2H	$-Cl$	104
Diuron	Dealkylation	$-NH \cdot CO \cdot NH \cdot CH_3$	$-Cl$	$-Cl$	105
		$-NH \cdot CO \cdot NH_2$	$-Cl$	$-Cl$	105
	Hydrolysis	$-NH_2$	$-Cl$	$-Cl$	105,106
Linuron	Dealkoxylation	$-NH \cdot CO \cdot NH \cdot CH_3$	$-Cl$	$-Cl$	107

trole, benzoylprop-ethyl, together with the closely related flamprop esters, chlorsulfuron, diclofop-methyl, difenzoquat, glyphosate, and picloram.

No degradation products appear to have been isolated from soils treated with amitrole, difenzoquat, or picloram.

Benzoylprop (**19**; R = −Cl), which is applied as the ethyl ester, and flamprop (**19**; R = −F) which is formulated as the methyl and *iso*-propyl esters, have the acid (**19**) as their basic structure.

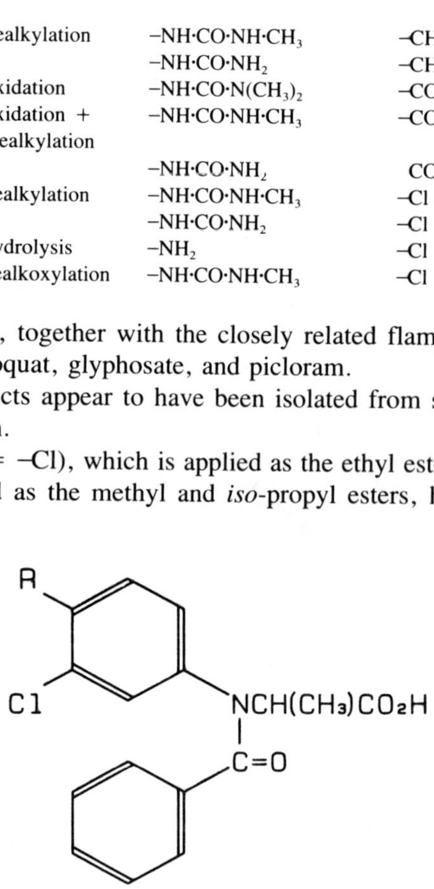

19

Under laboratory and field conditions, the esters of benzoylprop and flamprop undergo a slow hydrolysis in soil to their corresponding acids.[108-113] Studies with (^{14}C)benzoylprop acid (**19**; R = −Cl) have shown [108] that under laboratory conditions, further transformations can occur with loss of the benzoyl group to form 3,4-dichlorophenylalanine and 3,4-dichloroaniline. (^{14}C)Flamprop acid (**19**; R = −F) can behave similarly,[111,112] being transformed to such products as 3-chloro-4-fluorophenylalanine and 3-chloro-4-fluoroaniline. Both anilines are bound to soil organic matter and do not appear to exist in a free state.

Chlorsulfuron has been shown to undergo transformation in field soils to 2-chlorobenzenesulfonamide.[114]

Diclofop-methyl (**20**; R = $-CH(CH_3)COOCH_3$) is a herbicidal ester that is rapidly hydrolyzed in moist soils under laboratory and field conditions to the acid form (**20**; R = $-CH(CH_3)COOH$.[113,115-117]

20

Two further transformation mechanisms have been identified as occurring during the degradation of the herbicide in laboratory and field soils. The one involves a decarboxylation[116,117] to a phenetole (**20**; R = $-CH_2 \cdot CH_3$), and the second invokes an ether cleavage to a phenol (**20**; R = $-H$).[115-117] Laboratory experiments with (^{14}C)diclofop-methyl have indicated[115] that ring fission of both aromatic systems does not occur to any significant extent.

Glyphosate (**21**) undergoes biological degradation in moist, nonsterile soils under laboratory conditions to aminomethylphosphonic acid (**22**), which undergoes further breakdown to unidentified products.[118,119]

21   **22**

V. CONCLUSIONS

This chapter has summarized the methods available for studying transformations of herbicides in soils and has listed degradation products isolated from soil treatments of commonly used herbicides.

Where information is available, it appears that herbicidal transformations in field soils are very similar to those occurring in soils under laboratory conditions. Also, there would seem to be little difference in the breakdown pathways of herbicides in soils from different parts of the world.

However, it is apparent that considerably more is known concerning the transformations of some herbicides than others. Thus, transformation products derived from dinitroaniline, triazine, and urea herbicides are better documented than other classes. Another fact that emerges is that the transformations experienced by newer herbicides are better reported than are those of the older chemicals.

There are two main reasons for this. (1) Prior to 1960, there was no precedent for metabolic studies to be undertaken for herbicide registration. Thus, no incentive was provided to undertake such studies. Today, registration regulations exist in several countries so that this information must now be provided. (2) Since the technology for instrument development has increased, the detection and analysis of minute traces of compounds has become possible. Mass spectrometry has revolutionized this area. The synthesis of (^{14}C)-labeled herbicides has made the study of their degradation a relatively simple matter. There has been a greater willingness on the part of the manufacturers to provide labeled materials to independent government and university laboratories, and this has stimulated much research and many publications on the topic of herbicide transformations. However, there is a reluctance on the part of most manufacturers to publish their research findings; consequently, about 75% of the references mentioned in this chapter describe studies carried out by government and university scientists.

Future research should be directed towards a better understanding of how the older chemicals, for which there is little or no information, are transformed in the soil. This would also include further study as to how chlorinated anilines are incorporated into soil organic matter and the significance of such interactions. Anilines result in the soil transformation of several classes of herbicides, including those of the amides, carbamates, dinitroanilines, ureas, and from benzoylprop and flamprop, but their incorporation into organic matter is poorly understood. The final fate of the triazine herbicides in soil is unknown, and virtually nothing has been reported on the transformation of the soil-incorporated thiocarbamates.

Today, for statisfactory pest management, crops may receive successive treatments with one or more herbicides, as well as fungicides and insecticides for additional pest control. Although some studies have been initiated to compare the degradation rates of herbicides in the presence and absence of additional pesticides, no work has been reported as to whether the transformation pathways discussed in this chapter are modified by such treatments.

REFERENCES

1. **Guth, J. A.,** The study of transformations, in *Interactions between Herbicides and the Soil,* Hance, R. J., Ed., Academic Press, London, 1980, chap. 5.
2. **Bartha, R. and Pramer, D.,** Features of a flask and method for measuring the persistence and biological effects of pesticides in soil, *Soil Sci.,* 100, 68, 1965.
3. **Harvey, J.,** A simple method for evaluating soil breakdown of ^{14}C-pesticides under field conditions, *Residue Rev.,* 85, 149, 1983.
4. **Kaufman, D. D. and Kearney, P. C.,** Microbial transformations in the soil, in *Herbicides, Physiology, Biochemistry, Ecology,* Vol. 2., 2nd ed., Audus, L. J., Ed., Academic Press, New York, 1976, chap. 2.
5. **Crosby, D. G.,** Nonbiological degradation of herbicides in the soil, in *Herbicides, Physiology, Biochemistry, Ecology,* Vol. 2, 2nd ed., Audus L. J., Ed., Academic Press, New York, 1976, chap. 3.
6. **Alexander, M.,** Pesticides, in *Introduction to Soil Microbiology,* 2nd ed., John Wiley & Sons, New York, 1977, chap. 26.
7. **Leasure, J. K.,** The halogenated aliphatic acids, *J. Agric. Food Chem.,* 12, 40, 1964.
8. **Foy, C. L.,** The chlorinated aliphatic acids, in *Herbicides, Chemistry, Degradation and Mode of Action,* Vol. 1, 2nd ed., Kearney, P. C. and Kaufman, D. D., Eds., Marcel Dekker, New York, 1975, chap. 8.
9. **Jaworski, E. G.,** Chloracetamides, in *Herbicides, Chemistry, Degradation and Mode of Action,* Vol 1, 2nd ed., Kearney, P. C. and Kaufman, D. D., Eds., Marcel Dekker, New York, 1975, chap. 6.
10. **Kaufman, D. D. and Blake, J.,** Microbial degradation of several acetamide, acylanilide, carbamate, toluidine and urea pesticides, *Soil Biol. Biochem.,* 5, 297, 1973.
11. **Bartha, R. and Pramer, D.,** Pesticide transformation to aniline and azo compounds in soil, *Science,* 156, 1617, 1967.
12. **Burge, W. D.,** Microbial populations hydrolyzing propanil and accumulation of 3,4-dichloroaniline and 3,3′,4,4′-tetrachloroazobenzene in soils, *Soil Biol. Biochem.,* 4, 379, 1972.
13. **Chisaka, H. and Kearney, P. C.,** Metabolism of propanil in soils, *J. Agric. Food Chem.,* 18, 854, 1970.
14. **Kearney, P. C. and Plimmer, J. R.,** Metabolism of 3,4-dichloroaniline in soils, *J. Agric. Food Chem.,* 20, 584, 1972.

15. **Still, G. G. and Herrett, R. A.,** Methylcarbamates, carbanilates, and acylanilides, in *Herbicides, Chemistry, Degradation and Mode of Action,* Vol. 2, 2nd ed., Kearney, P. C. and Kaufman, D. D., Eds., Marcel Dekker, New York, 1976, chap. 12.
16. **Hsu, T. S. and Bartha, R.,** Interaction of pesticide-derived chloroaniline residues with soil organic matter, *Soil Sci.,* 116, 444, 1974.
17. **You, I. S., Jones, R. A., and Bartha, R.,** Evaluation of a chemically defined model for the attachment of 3,4-dichloroaniline to humus, *Bull. Environ. Contam. Toxicol.,* 29, 476, 1982.
18. **Kearney, P. C., Smith, R. J., Plimmer, J. R., and Guardia, F. S.,** Propanil and TCAB residues in rice soils, *Weed Sci.,* 18, 464, 1970.
19. **Carey, A. E., Yang, H. S. C., Wiersma, G. B., Tai, H., Maxey, R. A., and Dupuy, A. E.,** Residual concentrations of propanil, TCAB, and other pesticides in rice-growing soils in the United States, 1972, *Pestic. Monit. J.,* 14, 23, 1980.
20. **Hargrove, R. S. and Merkle, M. G.,** The loss of alachlor from soil, *Weed Sci.,* 19, 652, 1971.
21. **Yih, R. Y., Swithenbank, C., and McRae, D. H.,** Transformations of the herbicide N-(1,1-dimethylpropynyl)-3,5-dichlorobenzamide in soil, *Weed Sci.,* 18, 604, 1970.
22. **Yih, R. Y. and Swithenbank, C.,** Identification of metabolites of N-(1,1-dimethylpropynyl)3,5-dichlorobenzamide in soil and alfalfa, *J. Agric. Food Chem.,* 19, 314, 1971.
23. **MacRae, I. C. and Alexander, M. A.,** Microbial degradation of selected herbicides in soil, *J. Agric. Food Chem.,* 13, 72, 1965.
24. **Dewey, O. R., Lindsay, R. V., and Hartley, G. S.,** Biological destruction of 2,3,6-trichlorobenzoic acid in soils, *Nature,* 195, 1232, 1962.
25. **Grover, R. and Smith, A. E.,** Adsorption studies with the acid and dimethylamine forms of 2,4-D and dicamba, *Can. J. Soil Sci.,* 54, 179, 1974.
26. **Smith, A. E.,** Transformation of dicamba in Regina Heavy Clay, *J. Agric. Food Chem.,* 21, 708, 1973.
27. **Smith, A. E.,** Breakdown of the herbicide dicamba and its degradation product 3,6-dichlorosalicylic acid in prairie soils, *J. Agric. Food Chem.,* 22, 601, 1974.
28. **Smith, A. E. and Muir, D. C. G.,** Determination of extractable and nonextractable radioactivity from small field plots 45 and 95 weeks after treatment with (^{14}C)dicamba, (2,4-dichloro(^{14}C)phenoxy)acetic acid, (^{14}C)triallate, and (^{14}C)trifluralin, *J. Agric. Food Chem.,* 32, 588, 1984.
29. **Smith, A. E.,** An analytical procedure for bromoxynil and its octanoate in soils; persistence studies with bromoxynil octanoate in combination with other herbicides in soil, *Pestic. Sci.,* 11, 341, 1980.
30. **Beynon, K. I. and Wright, A. N.,** The fate of the herbicides chlorthiamid and dichlobenil in relation to residues in crops, soils, and animals, *Residue Rev.,* 43, 23, 1972.
31. **Briggs, G. G. and Dawson, J. E.,** Hydrolysis of 2,6-dichlorobenzonitrile in soils, *J. Agric. Food Chem.,* 18, 97, 1970.
32. **Smith, A. E.,** Degradtion of bromoxynil in Regina Heavy Clay, *Weed Res.,* 11, 276, 1971.
33. **Verloop, A.,** Fate of the herbicide dichlobenil in plants and soil in relation to its biological activity, *Residue Rev.,* 43, 55, 1972.
34. **Verloop, A. and Nimmo, W. B.,** Metabolism of dichlobenil in sandy soil, *Weed Res.,* 10, 65, 1970.
35. **Zaki, M. A., Taylor, H. F., and Wain, R. L.,** Studies with 3,5-diiodo-4-hydroxybenzonitrile (ioxynil) and related compounds in soils and plants, *Ann. Appl. Biol.,* 59, 481, 1967.
36. **Hsu, J. C. and Camper, N. D.,** Degradation of ioxynil to CO_2 in soil, *Pestic. Biochem. Physiol,* 5, 47, 1975.
37. **Khan, S. U. and Miller, S. R.,** Effects of repeated application of dichlobenil in a commercial apple orchard, *J. Agric. Food Chem.,* 30, 1246, 1982.
38. **Calderbank, A. and Slade, P.,** Diquat and paraquat, in *Herbicides, Chemistry, Degradation and Mode of Action,* Vol. 2, 2nd ed., Kearney, P. C. and Kaufman, D. D., Eds., Marcel Dekker, New York, 1976, chap. 10.
39. **Bartha, R.,** Fate of herbicide-derived chloroanilines in soil, *J. Agric. Food Chem.,* 19, 385, 1971.
40. **Bartha, R. and Pramer, D.,** Transformation of the herbicide methyl-N-(3,4-dichlorophenyl)carbamate (swep) in soil, *Bull. Environ. Contam. Toxicol.,* 4, 240, 1969.
41. **Smith, A. E. and Milward, L. J.,** Thin-layer chromatographic detection of the herbicide asulam in soils and the identification of sulphanilamide as a minor soil degradation product, *J. Chromatogr.,* 265, 378, 1983.
42. **Smith, A. E. and Walker, A.,** A quantitative study of asulam persistence in soils, *Pestic. Sci.,* 8, 449, 1977.
43. **Golab, T. and Althaus, W. A.,** Transformation of isopropalin in soil and plants, *Weed Sci.,* 23, 165, 1975.
44. **Golab, T., Althaus, W. A., and Wooten, H. L.,** Fate of (^{14}C)trifluralin in soil, *J. Agric. Food Chem.,* 27, 163, 1979.
45. **Golab, T., Bishop, C. E., Donoho, A. L., Manthey, J. A., and Zornes, L. L.,** Behaviour of ^{14}C oryzalin in soil and plants, *Pestic. Biochem.,* 5, 196, 1975.

46. **Probst, G. W., Golab, T., Herberg, R. J., Holzer, F. J., Parka, S. J., Van Der Schans, C., and Tepe, J. B.,** Fate of trifluralin in soils and plants, *J. Agric. Food Chem.,* 15, 592, 1967.

47. **Probst, G. W., Golab, T., and Wright, W. L.,** Dinitroanilines, in *Herbicides, Chemistry, Degradation and Mode of Action,* Vol. 1, 2nd ed., Kearney, P. C. and Kaufman, D. D., Eds., Marcel Dekker, New York, 1975, chap. 9.

48. **Kearney, P. C., Plimmer, J. R., Wheeler, W. B., and Konston, A.,** Persistence and metabolism of dinitroaniline herbicides in soils, *Pestic. Biochem.,* 6, 229, 1976.

49. **Kearney, P. C., Plimmer, J. R., Williams, V. P., Klingebiel, U. I., Isensee, A. R., Laanio, T. L., Stoltzenberg, G. E., and Zaylskie, R. G.,** Soil persistence and metabolism of *N-sec*-butyl-4-*tert*-butyl-2,6-dinitroaniline, *J. Agric. Food Chem.,* 22, 856, 1974.

50. **Wheeler, W. B., Stratton, G. D., Twilley, R. R., Ou, L. T., Carlson, D. A., and Davidson, J. M.,** Trifluralin degradation and binding in soil, *J. Agric. Food Chem.,* 27, 702, 1979.

51. **Golab, T., Herberg, R. J., Gramlich, J. V., Raun, A. P., and Probst, G. W.,** Fate of benefin in soils, plants, artificial rumen fluid, and the ruminant animal, *J. Agric. Food Chem.,* 18, 838, 1970.

52. **Smith, R. A., Belles, W. S., Shen, K. W., and Woods, W. G.,** The degradation of dinitramine $(N^3,N^3$-diethyl-2,4-dinitro-6-trifluoromethyl-*m*-phenylenediamine) in soil, *Pestic. Biochem. Physiol,* 3, 278, 1973.

53. **Matsunaka, S.,** Diphenyl ethers, in *Herbicides, Chemistry, Degradation, and Mode of Action,* Vol 2., 2nd ed., Kearney, P. C. and Kaufman, D. D., Eds., Marcel Dekker, New York, 1976, chap. 14.

54. **Leather, G. R. and Foy, C. L.,** Metabolism of bifenox in soil and plants, *Pestic. Biochem. Physiol,* 7, 437, 1977.

55. **Kaufman, D. D.,** Phenols, in *Herbicides, Chemistry, Degradation and Mode of Action,* Vol. 2, 2nd ed., Kearney, P. C. and Kaufman, D. D., Eds., Marcel Dekker, New York, 1976, Chap. 13.

56. **Burcar, P. J., Wershaw, R. L., Goldberg, M. C., and Khan, L.,** Gas chromatographic study of the behaviour of the iso octyl ester of 2,4-D under field conditions in North Park, Colorado, *Anal. Instrum.,* 4, 215, 1967.

57. **Grover, R.,** The adsorptive behaviour of acid and ester forms of 2,4-D on soils, *Weed Res.,* 13, 51, 1973.

58. **Smith, A. E.,** The hydrolysis of 2,4-dichlorophenoxyacetate esters to 2,4-dichlorophenoxyacetic acid in Saskatchewan soils, *Weed Res.,* 12, 364, 1972.

59. **McKone, C. E. and Hance, R. J.,** Determination of residues of 2,4,5-T in soil by gas chromatography of the *n*-butyl ester, *J. Chromatogr.,* 69, 204, 1972.

60. **Smith, A. E.,** The hydrolysis of herbicidal phenoxyalkanoic esters to phenoxyalkanoic acids in Saskatchewan soils, *Weed Res.,* 16, 19, 1976.

61. **Smith, A. E. and Hayden, B. J.,** Hydrolysis of MCPA esters and the persistence of MCPA in Saskatchewan soils, *Bull. Environ. Contam. Toxicol.,* 25, 369, 1980.

62. **Grover, R. and Smith, A. E.,** Adsorption studies with the acid and dimethylamine forms of 2,4-D and dicamba, *Can. J. Soil Sci.,* 54, 179, 1974.

63. **Foster, R. K. and McKercher, R. B.,** Laboratory incubation studies of chlorophenoxyacetic acids in Chernozemic soils, *Soil Biol. Biochem.,* 5, 333, 1973.

64. **Kunc, F. and Rybářová, J.,** Mineralization of carbon atoms of ^{14}C-2,4-D side chain and degradation ability of bacteria in soil, *Soil Biol. Biochem.,* 15, 141, 1983.

65. **Smith, A. E.,** Identification of 2,4-dichloroanisole and 2,4-dichlorophenol as soil degradation products of ring-labeled (^{14}C)2,4-D, *Bull. Environ. Contam. Toxicol.,* 34, 150, 1985.

66. **Smith, A. E.,** Identification of 4-chloro-2-methylphenol as a soil degradation product of ring-labeled (^{14}C)mecoprop, *Bull. Environ. Contam. Toxicol.,* 34, 656, 1985.

67. **McCall, P. J., Vrona, S. A., and Kelley, S. S.,** Fate of uniformly carbon-14 ring labeled 2,4,5-trichlorophenoxyacetic acid and 2,4-dichlorophenoxyacetic acid, *J. Agric. Food Chem.,* 29, 100, 1981.

68. **Gutenmann, W. H., Loos, M. A., Alexander, M., and Lisk, D. J.,** *Beta*-oxidation of phenoxyalkanoic acids in soil, *Soil Sci. Soc. Am. Proc.,* 28, 205, 1964.

69. **Smith, A. E. and Hayden, B. J.,** Relative persistence of MCPA, MCPB and mecoprop in Saskatchewan soils, and the identification of MCPA in MCPB-treated soils, *Weed Res.,* 21, 179, 1981.

70. **Stott, D. E., Martin, J. P., Focht, D. D., and Haider, K.,** Biodegradation, stabilization in humus, and incorporation into soil biomass of 2,4-D and chlorocatechol carbons, *Soil Sci. Soc. Am. J.,* 47, 66, 1983.

71. **Wilson, R. G. and Cheng, H. H.,** Fate of 2,4-D in a Naff silt loam soil, *J. Environ. Qual.,* 7, 281, 1978.

72. **Paasivirta, J., Sattar, M. A., Lahtipera, M., and Paukku, R.,** GC and MS analysis of MCPA and its two metabolites in environment nearby the brush killing treatment zone, *Chemosphere,* 12, 1277, 1983.

73. **Kaufman, D. D.,** Degradation of carbamate herbicides in soil, *J. Agric. Food Chem.,* 15, 582, 1967.

74. **Anderson, J. P. E. and Domsch, K. H.,** Microbial degradation of the thiolcarbamate herbicide, diallate, in soils and by pure cultures of soil microorganisms, *Arch. Environ. Contam. Toxicol.,* 4, 1, 1976.

75. **Anderson, J. P. E. and Domsch, K. H.,** Relationship between herbicide concentration and the rates of enzymatic degradation of ^{14}C-diallate and ^{14}C-triallate, *Arch. Environ. Contam. Toxicol.,* 9, 259, 1980.

76. **Anderson, J. P. E.,** Soil moisture and the rates of biodegradation of diallate and triallate, *Soil Biol. Biochem.,* 13, 155, 1981.

77. **Esser, H. O., Dupuis, G., Ebert, E., Marco, G. J., and Vogel, C.,** *s*-Triazines, in *Herbicides, Chemistry, Degradation and Mode of Action,* Vol. 1, 2nd ed., Kearney, P. C. and Kaufman, D. D., Eds., Marcel Dekker, New York, 1975, chap. 2.

78. **Kaufman, D. D. and Kearney, P. C.,** Microbial degradation of *s*-triazine herbicides, *Residue Rev.,* 32, 235, 1970.

79. **Rhodes, R. C.,** Soil studies with ^{14}C labeled hexazinone, *J. Agric. Food Chem.,* 28, 311, 1980.

80. **Webster, G. R. B. and Reimer, G. J.,** Field degradation of the herbicide metribuzin and its degradation products in a Manitoba sandy loam soil, *Weed Res.,* 16, 191, 1976.

81. **Harris, C. I.,** Fate of 2-chloro-*s*-triazine herbicides in soil, *J. Agric. Food Chem.,* 15, 157, 1967.

82. **Beynon, K.I., Stoydin, G., and Wright, A. N.,** A comparison of the breakdown of the triazine herbicides cyanazine, atrazine and simazine in soils and maize, *Pestic. Biochem. Physiol.,* 2, 153, 1972.

83. **Beynon, K. I., Stoydin, G., and Wright, A. N.,** The breakdown of the triazine herbicide cyanazine in soils and maize, *Pestic. Sci.,* 3, 293, 1972.

84. **Khan, S. U. and Hamilton, H. A.,** Extractable and bound (nonextractable) residues of prometryn and its metabolites in an organic soil, *J. Agric. Food Chem.,* 28, 126, 1980.

85. **Khan, S. U. and Ivarson, K. C.,** Microbiological release of unextracted (bound) residues from an organic soil treated with prometryn, *J. Agric. Food Chem.,* 29, 1301, 1981.

86. **Frank, R., Sirons, G. J., and Anderson, G. W.,** Atrazine: the impact of persistent residues in soil on susceptible crop species, *Can. J. Soil Sci.,* 63, 315, 1983.

87. **Khan, S. U. and Marriage, P. B.,** Residues of atrazine and its metabolites in an orchard soil and their uptake by oat plants, *J. Agric. Food Chem.,* 25, 1408, 1977.

88. **Khan, S. U. and Saidak, W. J.,** Residues of atrazine and its metabolites after prolonged usage, *Weed Res.,* 21, 9, 1981.

89. **Muir, D. C. G. and Baker, B. E.,** Detection of triazine herbicides and their degradation products in tile-drain water from fields under intensive corn (maize) production, *J. Agric. Food Chem.,* 24, 122, 1976.

90. **Muir, D. C. G. and Baker, B. E.,** The disappearance and movement of three triazine herbicides and several of their degradation products in soil under field conditions, *Weed Res.,* 18, 111, 1978.

91. **Sirons, G. J., Frank, R., and Sawyer, T.,** Residues of atrazine, cyanazine, and their phytotoxic metabolites in a clay loam soil, *J. Agric. Food Chem.,* 21, 1016, 1973.

92. **Khan, S. U., Marriage, P. B., and Hamill, A. S.,** Effects of atrazine treatment of a corn field using different application methods, times, and additives on the persistence of residues in soil and their uptake by oat plants, *J. Agric. Food Chem.,* 29, 216, 1981.

93. **Beynon, K. I., Bosio, P., and Elgar, K. E.,** The analysis of crops and soils for the triazine herbicide cyanazine and some of its degradation products, *Pestic. Sci.,* 3, 401, 1972.

94. **Yoo, J. Y., Muir, D. C. G., and Baker, B. E.,** Persistence and movement of cyanazine and procyazine under field conditions, *Can. J. Soil Sci.,* 61, 237, 1981.

95. **Khan, S. U. and Marriage, P. B.,** Residues of simazine and hydroxysimazine in an orchard soil, *Weed Sci.,* 27, 238, 1979.

96. **Gardiner, J. A., Rhodes, R. C., Adams, J. B., and Soboczenski, E. J.,** Synthesis and studies with 2-^{14}C-labeled bromacil and terbacil, *J. Agric. Food Chem.,* 17, 980, 1969.

97. **Wolf, D. C. and Martin, J. P.,** Microbial degradation of 2-carbon-14 bromacil and terbacil, *Soil Sci. Soc. Am. Proc.,* 38, 921, 1974.

98. **Geissbühler, H., Martin, H., and Voss, G.,** The substituted ureas, in *Herbicides, Chemistry, Degradation and Mode of Action,* Vol. 1, 2nd ed., Kearney, P. C. and Kaufman, D. D., Eds., Marcel Dekker, New York, 1975, chap. 3.

99. **Maier-Bode, H. and Härtel, K.,** Linuron and monolinuron, *Residue Rev.,* 77, 1, 1981.

100. **Savage, K. E.,** Adsorption and degradation of chlorbromuron in soil, *Weed Sci.,* 21, 416, 1973.

101. **Smith, A. E. and Briggs, G. G.,** The fate of the herbicide chlortoluron and its possible degradation products in soils, *Weed Res.,* 18, 1, 1978.

102. **Bozarth, G. A. and Funderburk, H. H.,** Degradation of fluometuron in sandy loam soil, *Weed Sci.,* 19, 691, 1971.

103. **Mudd, P. J., Hance, R. J., and Wright, S. J. L.,** The persistence and metabolism of isoproturon in soil, *Weed Res.,* 23, 239, 1983.

104. **Gross, D., Laanio, T., Dupuis, G., and Esser, H. O.,** The metabolic behaviour of chlortoluron in wheat and soil, *Pestic. Biochem. Physiol.,* 10, 49, 1979.

105. **Dalton, R. L., Evans, A. W., and Rhodes, R. C.,** Disappearance of diuron from cotton fields, *Weeds,* 14, 31, 1966.

106. **Khan, S. U., Marriage, P. B., and Saidak, W. J.,** Persistence and movement of diuron and 3,4-dichloroaniline in an orchard soil, *Weed Sci.,* 24, 583, 1976.

107. **Mapplebeck, L. and Waywell, C.,** Detection and degradation of linuron in organic soils, *Weed Sci.,* 31, 8, 1983.

108. **Beynon, K. I., Roberts, T. R., and Wright, A. N.,** Degradation of the herbicide benzoylprop-ethyl in soil, *Pestic. Sci.,* 5, 451, 1974.
109. **Bosio, P. G., Cole, E. R., Mathews, B. L., Woodbridge, A. P., and Wright, A. N.,** The determination and study of residues of flamprop-isopropyl in soil, *Pestic. Sci.,* 13, 63, 1982.
110. **Bosio, P. G., Elgar, K. E., Mathews, B. L., Woodbridge, A. P., and Wright, A. N.,** Depletion of residues of benzoylprop-ethyl in soil, *Pestic. Sci.,* 13, 1, 1982.
111. **Hitchings, E. J. and Roberts, T. R.,** Degradation of the herbicide flamprop-isopropyl in soil under laboratory conditions, *Pestic. Sci.,* 10, 1, 1979.
112. **Hitchings, E. J. and Roberts, T. R.,** Degradation of the herbicide flamprop-methyl in soil, *Pestic. Sci.,* 11, 591, 1980.
113. **Smith, A. E.,** Extraction of free and bound carboxylic acid residues from field soils treated with the herbicides, benzoylprop-ethyl, diclofop-methyl, and flamprop-methyl, *J. Agric. Food Chem.,* 27, 428, 1979.
114. **Smith, A. E. and Hsiao, A. I.,** Transformation and persistence of chlorsulfuron in prairie field soils, *Weed Sci.,* 33, 555, 1985.
115. **Martens, R.,** Degradation of the herbicide (^{14}C)-diclofop-methyl in soil under different conditions, *Pestic. Sci.,* 9, 127, 1978.
116. **Smith, A. E.,** Degradation of the herbicide dichlorfop-methyl (sic) in prairie soils, *J. Agric. Food Chem.,* 25, 893, 1977.
117. **Smith, A. E.,** Transformation of (^{14}C)diclofop-methyl in small field plots, *J. Agric. Food Chem.,* 27, 1145, 1979.
118. **Nomura, N. S. and Hilton, H. W.,** The adsorption and degradation of glyphosate in five Hawaiian sugarcane soils, *Weed Res.,* 17, 113, 1977.
119. **Rueppel, M. L., Brightwell, B. B., Schaefer, J., and Marvel, J. T.,** Metabolism and degradation of glyphosate in soil and water, *J. Agric. Food Chem.,* 25, 517, 1977.

INDEX

A

Acetochlor, 13
Acids, 175
Activation energies, for dissipation, 150—153,
  156—157, 164
ACTMO, see Agricultural chemical transport model
Adsorption
  correlation with bioavailability, 12—15
  kinetics of, 4, 50
  mechanisms of, 4—6
  methods for measuring, 2—3
  numerical description of, 3—4
  and pesticide transport, 50
  in residue redistribution, 92
  temperature effects on, 98
Adsorption—desorption relations, 50
Adsorption isotherm, 3, 50, 93, 99
Adsorption partition coefficients, 50
Aerodynamic method, 109—111
Aggregation, soil, 2—4, 49, 53
Agricultural chemical transport model (ACTMO), 72
Agricultural runoff model (ARM), 72
Alachlor
  adsorption—activity correlations for, 13
  availability index of, 71
  dissipation of, 138, 149—150, 160
  groundwater contamination with, 37
  persistence of, 136
  runoff losses of, 56—57, 60—61
  solubility and Henry's law coefficient of, 97
  structure and transformations of, 177—179
  vapor pressure of, 94
  volatilization rates of, 116, 118—120, 123
Alcohols, 175
Aldicarb, 31, 34
Aliphatic acids, 177
Ametryn, 150, 189—191
Amides, 175
  dissipation of, 138
  transformation reactions of, 177—179, 196
Amines, 175
Aminomethylphosphonic acid, 195
Aminotriazole, 5
Amitrole, 150, 155, 193—194
Anilines, 175, 183, 196
Anions, 5
Application factors, 54, 135
Application losses, 132
Aquatic vegetation, 66
Aquifer, 37—38
ARM , see Agricultural runoff model
Aromatic carboxylic acids, 163
Arrhenius equation, 156, 164
Asellus, 14
Asulam
  dissipation of, 138, 147, 150, 157
  structure and transformation of, 182—183
Atrazine
  adsorption and uptake of, 13—14
  aquatic vegetation, effects on, 66
  availability index of, 71
  concentrations of, 46, 49, 55—60
  dissipation of, 136—137, 140, 150, 159
  groundwater contamination with, 37
  leaching of, 31—35
  structure and transformation of, 189—191
  vapor pressure of, 94
  volatilization of, 116, 118—120, 122
Attenuation factor, 38
Availability classes, 161—162
Availability index, 70—71
Avena fatua, 10—12

B

Back-diffusion, 92, 104, 120
Barban, 182—183
Barley, 11
Bartha and Pramer flasks, 173
Batch method, 2—3
Benefin, 138, 145—146, 183—185
Bensulide, 138
Bentazon, 136
Benzoic acids, 138, 163, 180
Benzonitriles, 138, 180—182
Benzoylprop—ethyl, 194, 196
Bifenox, 184
Bioaccumulation, 90, 124
Bioavailability, correlation with adsorption, 12—15
Bioconcentration factors, 7, 135
Bioelimination, 145
Biological activity, 14
Biotic/abiotic transformations, 25
Bipyridinium compounds, 7, 12
Bipyridyls, 138
Black Creek watershed, 66
Boundary layer
  atmospheric, 92, 126
  turbulent, 93
Bowen ratio method, 110—111
Bromacil
  dissipation of, 137, 150
  distribution in chambers, 136
  groundwater contamination with, 37
  solubility and Henry's law coefficient of, 97
  structure and transformation of, 191—192
  vapor pressure of, 94
Bromoxynil, 138, 180—181
Buffer strips, vegetative, 54
Butachlor, 177—179
Butralin, 138, 146, 183—185
Butylate, 136, 140, 160

C

Capillary action, 91
Capillary flow, 106
Carbamates, 138, 182—183, 196
Carbaryl, 71

D

2,3,6-TBA, 180
TCA, 177
2,3,7,8-TCDD, 160
Temporal variability, 29
Terbacil, 141, 191—192
Terbuthylazine, 71
TFM, see Transfer function model
Thioalcohols, 175
Thiocarbamates
  dissipation of, 140, 155
  transformation reactions of, 188—189, 196
Tifton loamy sand, 75—80
Topological indices, molecular, 8
Toxaphene
  availability index of, 71
  dissipation of, 160
  rainfall washoff of, 51
Toxicity tests, 46
  See also Phytotoxicity
Transfer function model (TFM), 36
Transformations in soil
  methods for studying, 172—173
  reactions in, 173—176
Transpiration stream concentration factor, 9, 11
Triallate
  shoot uptake of, 10
  solubility and Henry's law coefficient of, 97
  structure and transformation of, 189
  vapor pressure of, 94
  volatilization of, 104—107
Triazines
  adsorption mechanisms of, 5—7
  dissipation of, 140—141
  lethal concentration of, 11
  rainfall washoff of, 51
  transformation reactions of, 189—191, 196
Tricamba, 180
2,4,5-Trichloroanisole, 188
Trichlorobenzoic acid (TBA), 163
2,4,5-Trichlorophenol, 188—
Trifluralin
  availability index of, 71
  desorption isotherms for, 97—98
  dissipation of, 137, 139, 159—160
  incorporation of, 92
  octanol—water partition values of, 9
  persistence of, 136, 146
  runoff losses of, 56—57, 62
  structure and transformation of, 183—186
  vapor pressure and solubility of, 94, 97, 100—102
  volatilization of, 104—105, 107, 112, 115—121

Turbulent flow, 47, 92, 126
Turnip, 12
Two-compartment exponential model, 144

U

Universal soil loss equation (USLE), 72, 78
Uracils, 141, 191—192
Ureas
  binding mechanism for, 5
  dissipation of, 141
  lethal concentration of, 11
  transformation reactions of, 192—193, 196
USLE, see Universal soil loss equation

V

Vadose zone, 31, 37
Van der Waals forces, 5—6
Vapor density, 92, 96—98
Vapor pressure, 93—96, 120
Variability, 29, 108
Vernolate, 140, 189
Viscosity, 92
Volatilization
  of herbicide, 90—92, 145, 154, 157—158, 164
  losses, factors in, 132
  pesticide depletion by, 55
  rates, measurement of, 107—123
  soil moisture content and, 99—102, 161
Vulnerability, aquifer, 37—38

W

Water
  bodies, herbicides in, 66
  extraction, 13
  movement in soils, 22—27
  solubility, 7
  surfaces, volatilization from, 123
Water vapor flux method, 109—111
Weather variables, 143
Wheat, 10—11
Wick effect, 106, 117—118, 120
Wind erosion, 122—123
Windspeed, 109—110
Wye River estuary, 66

X

XAD resin, 114